ART HACK PRACTICE

Bridging art and innovation, this book invites readers into the processes of artists, curators, cultural producers and historians who are working within new contexts that run parallel to or against the phenomenon of 'maker culture.'

The book is a fascinating and compelling resource for those interested in critical and interdisciplinary modes of practice that combine arts, technology and making. It presents international case studies that interrogate perceived distinctions between sites of artistic and economic production by brokering new ways of working between them. It also discusses the synergies and dissonances between art and maker culture, analyses the social and collaborative impact of maker spaces and reflects upon the ethos of the hackathon within the fabric of a media lab's working practices.

Art Hack Practice: Critical Intersections of Art, Innovation and the Maker Movement is essential reading for courses in art, design, new media, computer science, media studies and mass communications as well as those working to bring new forms of programming to museums, cultural venues, commercial ventures and interdisciplinary academic research centres.

Victoria Bradbury is a new media artist and researcher who creates interactive installations, virtual reality artworks and performances. She teaches interactive New Media at The University of North Carolina Asheville and holds a PhD from the CRUMB research group at the University of Sunderland and an MFA from Alfred University. Her website is www.victoriabradbury.com.

Suzy O'Hara is a curator and researcher based at the University of Sunderland. Through her practice, she interrogates evolving relationships between art and technology and investigates interdisciplinary and cross-industry models of curation that utilise and inform innovation-based strategies. She holds a PhD from the CRUMB research group at the University of Sunderland and an MA in Curatorial Studies from Newcastle University.

ART HACK PRACTICE

Critical Intersections of Art, Innovation and the Maker Movement

Edited by Victoria Bradbury and Suzy O'Hara

Routledge
Taylor & Francis Group
NEW YORK AND LONDON

First published 2020
by Routledge
52 Vanderbilt Avenue, New York, NY 10017

and by Routledge
2 Park Square, Milton Park, Abingdon, Oxon, OX14 4RN

Routledge is an imprint of the Taylor & Francis Group, an informa business

Library of Congress Cataloging-in-Publication Data
A catalog record for this title has been requested

ISBN: 978-0-8153-7492-3 (hbk)
ISBN: 978-0-8153-7491-6 (pbk)
ISBN: 978-1-351-24121-2 (ebk)

Typeset in Bembo
by codeMantra

to sally, ruby, xavi + zora

CONTENTS

FIGURES AND TABLE

Figures

Table

CONTRIBUTORS

Sally Abu Bakr holds a BA in Economics from Birzeit University. Since 2009, she has been part of the Ramallah Municipality, previously appointed as Art and Cultural Projects Supervisor, since 2016 she has been the Director of Culture and Social Affairs. For seven years, she managed the first public spaces art festival in Palestine, "Wein a Ramallah," and played a significant role since then in the development and strategic mapping of the art and culture program in the city of Ramallah as a whole. http://www.ramallah.ps/english.aspx

Tania Aedo is a cultural producer with a long trajectory developing projects at the intersections of knowledge, especially art, science and technology. She has been Director of Laboratorio Arte Alameda since 2007 and is former director of Centro Multimedia at the National Center for the Arts in México. She studied at the National Visual Art School and at Escuela Superior de Artes de Yucatán. Aedo has been honoured with grants such as the Media Arts Fellowship from foundations including Rockefeller, Ford and McArthur, the Creative Residencies Program at the Banff Center and from Fondo Nacional para la Cultura y las Artes.

Victoria Bradbury is a new media artist and researcher who creates interactive installations, virtual reality artworks and performances. She teaches interactive New Media at The University of North Carolina Asheville and holds a PhD from the CRUMB research group at the University of Sunderland and an MFA from Alfred University. Her website is www.victoriabradbury.com.

Serena Cangiano is a senior researcher at SUPSI University of Applied Sciences and Arts of Southern Switzerland, where she coordinates the MAS in

Interaction design and manages FabLab Lugano. She holds a PhD from University Iuav of Venice with a thesis on open source practices and interaction design. http://www.morethanhelvetica.ch

Ruth Catlow is an artist who works with emancipatory network cultures, practices and poetics. She is a co-founder and artistic director, with Marc Garrett, of Furtherfield for arts, technology and social change. Furtherfield's public exhibition and lab venues in the heart of Finsbury Park, London, provide a unique physical interface for free exhibitions, events and workshops, and an online hub provides a forum for exchange and critical review. These spaces bring together diverse local cultures with thriving networks of international artists, technologists and activists. She published *Artists Re:thinking the Blockchain* with Marc Garrett, Nathan Jones, and Sam Skinner in 2017.

Julie Freeman is an artist who translates complex processes and data from natural sources into kinetic sculptures, physical objects, images, sound compositions and animations. Her work explores relationships between science and the natural world; questioning the use of technology in how we translate nature. She often works collaboratively and experimentally with scientists. Julie is Art Associate at the Open Data Institute, a TED senior fellow and a co-founder of Fine Acts.

Davide Fornari is an associate professor at ECAL/University of Art and Design Lausanne, where he has led the R&D sector since 2016. Previously he was a teacher and a researcher at SUPSI University of Applied Sciences and Arts. He holds a PhD in Design Sciences from University Iuav of Venice. http://www.davidefornari.eu

Marc Garrett is co-director and co-founder, with artist Ruth Catlow, of the arts collective Furtherfield in Finsbury Park, London. He has curated over 50 contemporary media arts exhibitions and projects nationally and internationally. He curated the exhibition 'Monsters of the Machine: Frankenstein in the 21st Century,' at Laboral, Spain. He is the main editor of the Furtherfield website. He has written various books and articles about art, technology and social change. He published 'Artists Re:thinking the Blockchain' with Ruth Catlow, Nathan Jones and Sam Skinner in 2017. He is currently in the final year of his PhD at the University of London, Birkbeck College.

Beryl Graham is Professor of New Media Art at the University of Sunderland and is co-founder of the CRUMB resource for curators of New Media Art. She curated the international exhibition *Serious Games* for the Laing and Barbican art galleries and has also worked with The Exploratorium, San Francisco, and *San Francisco Camerawork*. Her books include *Digital Media Art* (Heinemann 2003),

Rethinking Curating: Art After New Media (MIT Press 2010 with Sarah Cook) and *New Collecting: Exhibiting and Audiences* (Ashgate 2014 ed.). She has written for periodicals including *Art Monthly, Leonardo* and the *Journal of Curatorial Studies.* She has been an invited speaker at conferences including *Navigating Intelligence* (Banff Centre for the Arts), *Creativity and Cognition* (Sydney), *IRL to URL* (BALTIC Centre for Contemporary Art), *Decoding the Digital* (Victoria and Albert Museum) and *Cultural Value and the Digital* (Tate Modern). http://www.berylgraham.com

Jasmin Theresa Grimm is a curator, lecturer and creative producer for international cultural projects. With a background in 'Communication in Social and Economic contexts' (MA, University of the Arts Berlin), she develops programmes such as workshops, festivals and exhibitions at the intersection of art, society and technology. She is a member of the Robert Bosch Cultural Manager Network, Tandem Shaml – Cultural Manager Exchange Alumni and the Cultural Innovators Network between Europe and the MENA region. www.jasmingrimm.com

Olof Mathé has degrees in physics and philosophy from the Royal Institute of Technology, Stockholm, and École normale supérieure, Paris. He created Art Hack Day, a global computer arts hackathon. He is currently the co-founder & CEO of Mixmax, reinventing email for the 21st century.

Alexia Mellor is an interdisciplinary artist and researcher exploring how participatory artistic practice might serve to create a critical, reflective space to examine global 'wicked problems' and the relevance of these issues within local contexts. Mellor places particular focus on using infrastructuring to create platforms for transforming participants into active citizens. She has exhibited and conducted site-responsive projects internationally, recently completing an AHRC-funded placement collaborating with Fab Lab Barcelona and the H2020 Making Sense project. She completed her PhD at Newcastle University in 2019. http://alexiamellor.com

Olga Mink is the director of Baltan Laboratories, which initiates innovative research and development at the intersections of disciplines. Olga studied Fine Art and Animation Design and obtained her Master's at the Sandberg Institute in Amsterdam. She worked as a media artist and independent curator for media festivals. Olga co-founded Age of Wonderland, a four-year exchange program to boost social innovation together with the NGO Hivos during Dutch Design Week. Olga is the chair of Trudo.lab for the social housing corporation Sint Trudo. She is also affiliated as a researcher for "Design for Transformative Practices" at Fontys University of Applied Sciences. http://baltanlaboratories.org

Mugendi K. M'Rithaa is a transdisciplinary industrial designer/educator/researcher and a committed advocate of Socially Conscious Design who has a special interest in advancing the developmental agenda in Afrika. He has been a curator for industrial design for the Design Indaba Expo since 2014 and is associated with a number of international networks focusing on design within majority world contexts. He is President Emeritus of the World Design Organization supporting the aspirations of younger designers worldwide.

Ayodamola Tanimowo Okunseinde (ayo) is a Nigerian-American artist and interactive designer living and working in New York. His works range from painting and speculative design to physically interactive works, wearable technology and explorations of "Reclamation." Okunseinde was the co-founder and creative director of Dissident Display Studios in Washington DC. He has exhibited and presented at the 11th Shanghai Biennale, Tribeca Storyscapes, EYEO Festival, Brooklyn Museum, M.I.T. Beyond the Cradle, and Afrotectopia amongst others. Okunseinde holds an MFA in Design and Technology from Parsons School of Design in New York where he is currently an adjunct faculty member.

Suzy O'Hara is a curator and researcher based at University of Sunderland. Through her practice, she interrogates evolving relationships between art and technology and investigates interdisciplinary and cross-industry models of curation that utilise and inform innovation-based strategies. She holds a PhD from the CRUMB research group at the University of Sunderland and an MA in Curatorial Studies from Newcastle University.

Nora O. Murchú is a curator and designer based in Ireland. Her practice explores how complex socio-technical systems are imagined, built and used. She has held positions as a research associate for the Interaction Research Studio at Goldsmiths, and CRUMB at the University of Sunderland, and curated exhibitions for institutions including the Science Gallery, Rua Red and Transfer Gallery. She is a Lecturer in Design at the University of Limerick.

Irini Papadimitriou is a curator, producer and cultural manager, working at the forefront of digital culture. As Digital Programmes Manager at the Victoria & Albert Museum, she initiated and was responsible for programmes such as the annual Digital Design Weekend festival and Digital Futures, among other projects. She was also Head of New Media Arts Development at Watermans, exploring digital culture from a critical perspective through her programme. She has been co-curating the Artist Open Web platform at MozFest, and she is a co-founder of Maker Assembly. In October 2018, Irini moved to the position of Creative Director at FutureEverything.

Ellen Pearlman is a New York-based new media artist, writer, critic and curator who received her PhD from the School of Creative Media, Hong Kong City University. A Fulbright World Learning Specialist in Art, New Media and Technology, Ellen is on the faculty at Parsons/New School University. She is Director of the ThoughtWorks Arts Residency, President of Art-A-Hack (TM), and Director of The Volumetric Society of New York. She directed and created "Noor: A Brain Opera," the world's first immersive, interactive brain opera in a 360-degree theatre. She has presented her work and lectured globally. https://epmexico.wixsite.com/ellenpearlman

Clare Reddington joined Watershed in 2004, establishing creative technology programmes including Pervasive Media Studio and Playable City. She became Creative Director in 2014, looking after the Studio, engagement and cinema teams, and CEO in 2018. Clare works with industry, academic and creative partners from around the world to champion inclusion, support talent and develop new ideas. Clare was Executive Producer of REACT, a Visiting Professor at University of the West of England. She is chair of Emma Rice's Wise Children, and on the boards of RSC and Theatre Bristol.

Hannah Redler Hawes is an independent curator/producer specialising in contemporary art, new media and participation. She works with researchers from across the arts, sciences, technology and emerging fields to develop projects that explore shared interests, diversion and conversation between disciplines for broad audiences in galleries, museums, private organisations and digital spaces.

Azalea Seratoni is a Milan-based art curator, critic and historian of art. She is adjunct professor at Scuola Politecnica di Design (SPD), Milan and Istituto Europeo di Design (IED), Milan. She graduated in Art History, Faculty of Letters and Philosophy at Università degli Studi di Milano.

Abigail Susik is Associate Professor of Art History at Willamette University who writes about modern and contemporary art and aesthetics. She was a member of the New Media Caucus Board of Directors and an Associate Editor of its journal, *Media*-N. Susik was a guest co-editor of *Media*-N's Fall 2016 issue, "Uncovering News: Reporting and Forms of New Media Art" and has published numerous essays and reviews on subjects related to new media art.

Yidi Tsao works in the realm of contemporary art as writer, curator, artist and translator. Her research interests include the creative application of media technology, free culture movement, hacker/maker spaces and the role of art in community building. She is well versed in Chinese and Western culture and society while working towards bridging the gaps and promoting mutual understanding. She currently lives between Berlin and Hong Kong.

FOREWORD

Rethinking Art Hacks for the Future

Beryl Graham

> … the values are created by the people that we work with, you can't jump in with a template […] you have to experience it to know what it is.
>
> Marc Garrett of Furtherfield, London, during *Curating Art After New Media* short course

A long time ago, in a Britain far, far away, I was curating an exhibition of participatory art, titled *Serious Games*, for two art galleries/museums. At one point both organisations had doubts about one word in the title, and my challenge was in resolving the interesting fact that they objected to *different* words, and hence revealed the problematic boundaries between them. In putting 'art' and 'hack' in the title of this book, Bradbury and O'Hara have bravely challenged some borders of their own, as even the arts within themselves are currently removing some bricks from the hierarchical wall between craft, design, participatory practice and art. This approach, to me, seems entirely fitting to the exciting nature of new media, and to the interesting times in which we live.

This tension between territories, disciplines and sectors is way too important to be resolved in purely academic ways, and I'm very impressed at the editors' determination to select authors who are leading in the field right now, whether as makers, critics, activists, curators or theorists. This means that *how* to art hack is very thoroughly covered, as well as *why*. I'm hence very proud to see their rigorous doctoral research applied and communicated in such an accessible, useful and lively book. Despite much current rhetoric about cross-disciplinary research and cross-sector production, these things are much easier said than done. This book has clearly succeeded in doing so: critically pinning down the rhetorics into defined arguments that can be translated between fields, and in respecting the values of the very different disciplines. Admirably, readers from the arts need not fear being faced with uncritical hyperbole about Blockchain, and coders need not fear uninformed meanderings about Open Source.

These authors are also well informed on politico-economic contexts: the where and the who of art hacks. As Mugundi K. M'Rithaa points out, the "manufacturing base" of a country is inextricably linked to all kinds of systems of making and is of course very variable in different countries. I am writing this in the UK, where manufacturing or making skills are inextricably related to values of class, and specifically, in post-industrial Newcastle, where the remaining heavy engineering skills have had to become increasingly specialist (oil rigs and pedestrian bridges for example). In visiting China, India and South Africa, I have been struck by the decidedly different values attached to manufacturing, the very visible making skills, and the inventive flexibility of changing systems to make things work, from paying individuals to use their mobile phone for a call, through mapping DIY border crossings, to hacking electrical cables. These hacked systems are joyfully different to bland 'globalisation' and can be applied to social systems, electronic systems and to specific local production such as Palestinian ceramics.

In appreciating these hacked systems as possible futures, I'm keen to avoid the relentlessly futuristic discourses of speedy technological 'progress.' As Irini Papadimitriou points out, museums tend to change rather slowly, which helps them avoid foolishly half-baked early adoption of newer technologies. Like this book, I'm much more interested in what skills people might need for the future, and as an educator I've been lucky enough to see these art hack skills illuminate the practice of people coming from fields as diverse as engineering, activism, face-painting, coding and art history.

The opening quote from Marc Garrett, co-director of Furtherfield (the arts, technology and social change organisation) comes from a visit during the one-week short course *Curating Art After New Media*. By visiting places of production such as Machines Room, the course aims to add to the understanding of new media systems and hence enable participants to be able to change them in the future. Knowledge has been shared by attendees from India, Hong Kong, Bahrain, USA, Canada and Europe, and even traditional museum curators have been able to affect their organisations' systems of production, exhibition, collection, repository, education and distribution. This book presents such a gleeful array of inspirational modes of art hacking, whether the participants are jesters, prototypers or Afronauts, that I'm cheered at the prospect of creatively hacking our way into the future.

ACKNOWLEDGEMENTS

We would like to thank our colleagues at University of North Carolina Asheville and the University of Sunderland who encouraged us and gave us the space to conduct this research, especially Beryl Graham, Arabella Plouviez, Curt Cloninger, Lei Han, Christopher Oakley and Mark Hursty. Thanks to the students in the UNC Asheville New Media department who always inspire and welcome new ideas and to the UNCA Division of Natural Sciences and University Research Council for their financial support. Thanks to Creative Fuse North East, for supporting the exploration of the ways in which ideas and approaches from creative practice can stimulate innovation in many other sectors and to the Faculty of Arts and Creative Industries, University of Sunderland, for its financial support.

We are greatly indebted to CRUMB, the Curatorial Resource for Upstart Media Bliss, where we forged our friendship. Thanks also to New Media Caucus who provided us the opportunity to host our Reframing Innovation panel at CAA NYC, 2019.

We appreciate colleagues far and wide who influence this work, including xtine burrough who offered essential support and encouragement at the outset of our proposal, Paul Catanese who read drafts and offered feedback, as well as Richard Clay, Beryl Graham and Sarah Cook who were so generous with their expertise and advise during the process.

A huge thanks to the artists and curators who have contributed so generously. Our conversations shaped what this book would become. This book was made possible by Routledge and our editor, Erica Wetter and her assistant Emma Sherriff.

Finally, thanks to our families, Mark, Tibor, Sally, Ruby, Lilla, Xavier and Zoravela (who arrived during this process), Mary, Emily and Tom for their continued support and to Chris who is always in our thoughts.

+ CONTEXT 0

https://whereby.com/arthackpractice

INTRODUCTION

Victoria Bradbury and Suzy O'Hara

Art hacking is a term that seeks to describe how pervasive technologies continue to disrupt traditional and hierarchical boundaries between the arts, innovation and society. *Art Hack Practice* presents an emerging dynamic arts ecology from a range of voices across four continents and ten countries. By inviting our contributors to share their projects and insights in the first person, we begin to get a sense of the shifting roles, developing working practices and inherent value systems within these new modalities of interdisciplinary creative practice.

The website https://whereby.com/arthackpractice became our virtual meeting room and familiar common space while developing this book. It is a URL that defies disciplines, sectors, industries and locations by functioning as a space for face-to-face conversation between geographically and culturally diverse individuals and organizations. It is both private and public and has become a valued context where we ask our guest authors (and soon, you the reader), *how and why they do what they do.*

Art Hack Practice showcases artistic strategies that have emerged in response to an evolution of new contexts, highlighted as the section titles of this book. These include tangible infrastructures such as maker spaces, university fab labs, office spaces and media art festivals in traditional museums. Also included in this taxonomy are virtual spaces, such as within the canon of art history and within temporal communities. Many of these contexts are not created with the arts in mind, but artists, curators and historians continue to find strategies for leveraging them to accomplish projects that are hybrid in nature.

Art Hack Practice offers new insight into how traditional boundaries are shifting so that art can successfully operate within and alongside other sectors and industries such as open innovation, commercial digital industries and the natural sciences. These chapters include artists, designers, curators and historians who are working within, around or against the phenomenon known as 'maker

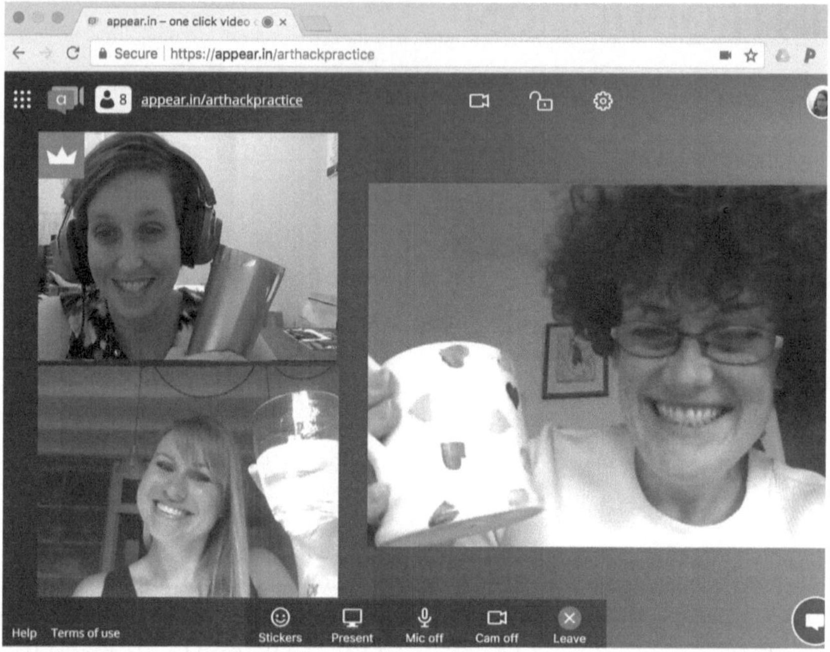

FIGURE I.1 Victoria Bradbury [top left] (Asheville, NC), Suzy O'Hara [right] (Newcastle, UK) and Alexia Mellor [bottom left] (Barcelona, Spain) at https://whereby.com/arthackpractice, where we met each of the authors as they developed their chapters.

culture.'[1] Each contributor provokes pertinent questions for us to consider: do these forms of practice suggest a modified 21st-century version of what artistic practices can be? In the context of 'The New Industrial Revolution,'[2] what novel artistic forms have emerged and how do they co-exist with established modes?

Generosity and inclusivity are at times purported as primary drivers of maker culture. In maker spaces, we are expected to 'share' and 'open source' our discoveries, just as the space shares tools amongst members. Similarly, this book resulted from the underlying generosity of our contributors who have openly shared their processes and reflections with you, providing valuable insights to seed future and ongoing projects and pedagogies.

Democratizing digital fabrication tools, technologies and making skills that have been relatively unavailable to most people, within inclusive and predominantly informal spaces, has catalyzed a global response that catapults making into mainstream consciousness. However, while there may be clear synergies in relation to generosity, inclusivity, resources and tools, the artists, hobbyists and commercial players who are found working across maker contexts often have differing aims, priorities, working practices and outputs.

Maker spaces are generally understood to be open environments that benefit members of their community with social and/or professional opportunities, and champion entrepreneurship, innovation and a capacity to enable marginalized groups to participate. However, as Sarah R. Davies suggests, while the motivations behind the Maker Movement[3] are laudable and well articulated, they are not without their challenges; who is included and excluded in maker communities is, in fact, always in question. Davies' research finds that while community is at the heart of the Maker Movement, many maker contexts have significant problems with inclusivity.[4] As such, there is a risk that intersections and clashes of worldviews may cause forms of dissonance that go unacknowledged and unnoticed by the more techno-positivist, innovation-focused voices coalescing around making and hacking.

Hacking:

> the activity of illegally using a computer to access information stored on another computer system or to spread a computer virus[5]

Within the framework of this widely known definition of hacking, a complex, abstract and globally distributed hacker class has emerged.[6] Hackers' activities have led to major consequences across the world, leading to a public perception of the 'hacker' as a nefarious criminal mastermind who can infiltrate networks, banking, personal communications and governments through black-boxed computer systems.

A more techno-solutionist[7] definition of the term hacking is manifest through maker culture. Within maker culture, is understood not as a nefarious act but rather as "a creative engagement with technologies"[8] through making. Within the context of a knowledge-led rather than a manufacturing-led economy, the Maker Movement is viewed as having significant promise for increasing social, economic and environmental sustainability.[9] By providing citizens with access to making tools and technologies, the hope is that personal and DIY fabrication will lead to the prosumption of new products, services and ideas that could go to market while promoting entrepreneurship.[10]

The impact of conflating 'hacking' with 'making' has been subject to criticism by Garnet Hertz, editor of Critical, who highlights that:

> "It's as if 'hacking' has been sanitized and transformed into 'making' – with politics, activism, tactics, history, economics and social issues removed in the process." Garnet Hertz, "Critical Making"[11]

Artists open up dialogues for a more critical understanding of the implications of how technologies are being used, particularly focusing on human aspects of technology.[12] By infusing maker culture with a multiplicity of artistic perspectives, a critical agenda can be reinstated. *Art Hack Practice* refocuses attention from maker contexts as sites for commercial ventures and startups to a mechanism for artists and curators to create and disseminate projects. By brokering new, direct ways of working between spaces of artistic and economic production, our authors challenge perceived distinctions between them.

There is a rich history of strategic programs that have enabled artists to transcend traditional borders with the arts and applied research contexts. Examples include Experiments in Art and Technology (E.A.T.),[13] the Xerox Parc PAIR program[14] and the Artist Placement Group.[15] Indeed, as media artist and educator Golin Levin points out, many contemporary products and services were initially developed within these cross-industry collaborations:

> ...some of today's most commonplace and widely-appreciated technologies were initially conceived and prototyped, years ago, by new-media artists.[16]

It should be highlighted, however, that these ideas were often appropriated and used commercially with no acknowledgement or financial gain for the artists involved in their production. For example, Michael Naimark created his work, Aspen Movie Map (1978–1980), the precursor to Google Maps[17] at MIT (with military funding from ARPA).

Today, there are many examples of programs and projects currently being delivered by organizations, artists and curators that enable engagement between the arts and maker, technology and commercially-driven cultures and resources. For example, Rhizome's Seven on Seven 'pairs seven leading artists with seven visionary technologists, and challenges them to make something new.'[18] This conference-based event has, for the past ten years, opened up a valuable space for critical conversations that foreground fundamental issues at the intersection of art and technology. LEAN Artists is a 'Seed Accelerator for Artists' devised by Jeremy Bailey in 2016. The program provided funding and mentorship for 'early stage culture' to create 'culturally disruptive start-ups.'[19] Sunderland 10 × 10[20] commissioned ten artists and creatives based in the North East of England to develop new work in collaboration with ten local businesses. By fostering these new relationships and connections, Sunderland 10 × 10 provided a valuable opportunity for artists to learn from and share skills, ideas and knowledge with commercial businesses. The project proved to be an ambitious provocation that made a clear case for creative, arts-led intervention and artistic engagement as an innovation strategy for business and for business to be seen as a developing context in which artists can practice.

Though this kind of collaborative work has never been central to mainstream arts discourse, artists have worked collaboratively in groups of various formats and across disciplines throughout history.[21] It is therefore no surprise that artists, who have long been interrogating technological tools and their relationship to society, are continuing to do so in applied, alternative contexts today.

Practice is about doing as theorizing, showing rather than telling. The critical position of *Art Hack Practice* is complex because of the viewpoints of those invited to contribute. Some authors and projects co-exist within maker and commercial contexts. Many have created their own spaces or methods because they didn't see an existing model that fit their aims. Some describe their practices as innovation while others are critical of the momentum and reach of commercial technologies. These dissonances cause authors in the book, at times, to oppose one another's points of view. And of course, you the reader are invited to construct your own classification scheme as you work through these chapters.

Context I, *Histories and Futures,* relates today's emerging art historical and cultural practices to the social, technological and political developments of the times in which artists, curators and art groups worked. It also looks toward the future at ways in which practices may develop in response to existing formats. Abigail Susik is hacking art history by reflecting on the 20th-century avant-garde from a 21st-century perspective. Mugendi K M'Rithaa applies the 'k' in Afrika to offer an internal understanding of the continent versus the external understanding of Africa, then maps this view onto ways in which the notion of 'making' is embodied there in the past, present and future. Serena Cangiano, Davide Fornari and Azalea Seratoni bring members of artist collective Gruppo T, who were influential in the 1950s/60s Italian avant-garde, into a fab lab to work with young artists, designers and technologists to reimagine previous projects that critically engage with open source methods. Ruth Catlow and Marc Garrett had been 'hacking' what an arts organization could be long before the term 'art hack' was coined. In Chapter 4, they examine new media art histories, particularly Net Art, while underlining current projects that support artists who are employing blockchain technologies. These projects consider how a commerce-oriented tool is developing and how artists can represent it before its pervasiveness is all-consuming (as the internet has become).

Labs and Fab Labs, Context II, examines the physical infrastructure of different types of city and university labs and organizations that bridge maker culture, the arts and broader contexts. These spaces interrogate issues around financial sustainability, staffing and office culture. In Chapter 5, Clare Reddington of Bristol UK's Pervasive Media Studios discusses the growth of the studio during its first ten years. The concept for the lab grew out of the question of how a hack can be extended into a physical space that supports the kinds of practices that would occur in a more temporary space. What the founders kept and left out and how this evolved over time – including folding social etiquette such as interruptibility, openness and generosity into a formalized contract that sustains the lab environment. Olga Mink, Director of Baltan Labs in Eindhoven, Netherlands, discusses ways in which Baltan hacks systems of open and social innovation while redefining the role of the artist within those modes. Notions of free experimentation sit alongside fundamental research as methods for brokering artists into broader conversations. Alexia Mellor reflects on the strengths that participatory artists can bring to technological projects implemented by fab labs in urban communities. Mellor's chapter places people, rather than technologies, at the center.

Context III, *Engaged Communities*, features methods in which participants engage with a fusion of arts and making. Arts are demonstrated to bridge between publics and innovation-focused spaces. While the tech community might state that the public is welcome to these spaces,[22] the point of entry is often unclear. This section fosters an understanding of technologies as they relate to the everyday but also to groups who aren't traditionally or widely represented in mainstream tech culture. Ayo Okunseinde frames how the Iyapo Repository project has developed across a residency at Eyebeam and through the Laundromat Project. He discusses sensitivities that artists encounter as they work between technology and participants, particularly communities of color, while imagining, fabricating and exhibiting artworks. In Chapter 9, Jasmin Theresa Grimm and Sally Abu Bakr connect heritage, craft, art and technology through the political context of Palestinian territory. In doing so, they bridge across time while crossing borders that can be physical, cultural or technological. Suzy O'Hara discusses Little Inventors, a project that was originally seeded within an artist's practice which then developed into a global creative brand. It exemplifies the value of nurturing creative confidence and skills within children. Little Inventors challenges preconceptions around the value of childrens' ideas by taking them seriously and connecting them with the professional skills of makers to turn them into real objects that have the potential to shape our future.

In Context IV, *Hack Events, Residences and Workshops*, temporary infrastructures of the Maker Movement and more traditional arts infrastructures are straddled by a series of practitioners. Ellen Pearlman accesses Actor Network Theory as a pedagogical strategy that benefits Parsons School of Design students alongside artist-participants in the Cyborg Arts Co-Lab. Victoria Bradbury and Suzy O'Hara discuss the hack event as activated within the arts by considering the background of participants and the importance that context brings to an event. Victoria Bradbury speaks with Constant Dullaart about ways in which artists find resources and places to make technological projects. In Chapter 14, Olof Mathé discusses Art Hack Day, a series of curated art hacks that experiment with the format of a 'traditional' tech sector hack in order to enable and resource artists to catalyze new collaborations and create art that interrogates technology.

Context V, *Museums, Galleries, Festivals and Programs*, showcases evolving curatorial modes. While curating is traditionally focused on caring for collections, here, curators are hacking spaces, practices, funding strands and other arts and non-arts infrastructures that can offer support to projects. In doing so, the curator seeks out fringe methods while mediating between powerful stakeholders, artists and creative practices. In Chapter 15, Nora O'Murchú hacks traditional curatorial models by involving her experience as a coder and maker herself into the curatorial process. Yidi Tsao reflects on the WIKITOPIA festival that hacks the economic and political infrastructure of Hong Kong to make space for artistic practices. Tsao centralizes the role of Ellen Pau, the 'godmother' of the media art scene in Hong Kong, who has developed ways to support artists in a city with limited space and skyrocketing costs of real estate. Julie Freeman and Hannah Redler Hawes hack the Open Data Institute office and coworking research space with artworks and curatorial programming. Tania Aedo's programming at Laboratorio Arte Alameda activates a historic church in Mexico City with interdisciplinary arts projects that reflect notions of perception, language and the multitude of cultures that co-exist within a giant metropolis. Irini Papadimitriou discusses two distinct but linked programs developed for the Victoria and Albert Museum in London, Digital Futures and the annual Digital Design Weekend. By enabling sharing, collaboration and exchange, Papadimitriou rethinks the role of the museum in the 21st century as a site for engaging in critical conversations about technology.

+ You

And so, in the spirit of generosity, we open this URL to you, the reader of this book. We invite you to see https://whereby.com/arthackpractice as a place to contribute and connect, co-work or co-hack, where you can share your own projects and initiatives with a growing community who are creating and supporting the art of our time. Open up your browser, type in our URL and tell us … how do you define *art hack practice*?

Notes

1 The Maker Movement defined by Make Magazine is "a tech-influenced DIY community" https://makerfaire.com/maker-movement/. It is a phenomenon that came about because of a combination of technological advancement including microcontrollers and 3-D printing, public access to technical information through the internet and a system of permanent and temporary physical spaces, such as fab labs, maker spaces and Maker Faires in which people can meet, build and work on technical projects. It has also emerged as a reaction to a decline in manufacturing in Western countries. Fallows, James. "Why the Maker Movement Matters." *The Atlantic*, June 5, 2016. https://www.theatlantic.com/business/archive/2016/06/why-the-maker-movement-matters-part-1-the-tools-revolution/485720/.
2 Anderson, Chris. 2014. *Makers: The New Industrial Revolution*. New York: Random House USA.
3 Fallows, *The Maker Movement* (Note 1).
4 Davies, Sarah R. 2017. *Hackerspaces: Making the Maker Movement*. Cambridge, UK ; Malden, MA: Polity, 100.

5 "Hacking." n.d. *Cambridge Dictionary*. Accessed December 20, 2018. https://dictionary.cambridge.org/dictionary/english/hacking.
6 Wark, McKenzie. 2004. *A Hacker Manifesto*. Cambridge, MA: Harvard University Press.
7 Lindtner, Silvia, Shaowen Bardzell, and Jeffrey Bardzell. 2016. "Reconstituting the Utopian Vision of Making: HCI After Technosolutionism." In *Proceedings of the 2016 CHI Conference on Human Factors in Computing Systems*, 1390–1402.
8 Richterich, Annika, and Karin Wenz. 2017. "Introduction. Making and Hacking." *Digital Culture and Society* 3 (1): 5–22.
9 Millard, Jeremy, Marie Sorivelle, Sarah Deljanin, Elisabeth Unterfrauner, and Christian Voigt. 2018. "Is the Maker Movement Contributing to Sustainability?" *Sustainability* 10: 2212.
10 Ritzer, George. 2012. "'Makers' Are Better Seen as Prosumers." *George Ritzer* (blog). November 4, 2012. https://georgeritzer.wordpress.com/2012/11/04/makers-are-better-seen-as-prosumers/.
11 Hertz, Garnet. 2012. *Critical Making*. http://www.conceptlab.com/criticalmaking/.
12 Artworks that reflect upon the human use of geological resources are discussed in: Parikka, Jussi. 2015. *A Geology of Media*. Electronic Mediations, volume 46. Minneapolis; London: University of Minnesota Press.
13 Media Art Net. 2019. "Media Art Net | E.A.T. – Experiments in Art and Technology: Experiments in Art and Technology – Documents." *Text*. March 19, 2019. http://www.medienkunstnetz.de/works/documents/.
14 Harris, Craig (ed.). 1999. *Art and Innovation: The Xerox PARC Artist-in-Residence Program*. Leonardo Books. Cambridge, MA: MIT Press.
15 James, Sarah. n.d. "Artist Placement Group." *Frieze*. Accessed March 19, 2019. https://frieze.com/article/artist-placement-group.
16 Golan Levin and. Collaborators. "Flong Blog + News » New Media Artworks: Prequels to Everyday Life." Accessed March 19, 2019. http://www.flong.com/blog/2009/new-media-artworks-prequels-to-everyday-life/.
17 Naimark, Michael. n.d. "Aspen Moviemap." Accessed March 19, 2019. http://www.naimark.net/projects/aspen.html.
18 "Seven on Seven." n.d. Rhizome. Accessed March 19, 2019. http://rhizome.org/sevenonseven/.
19 Bailey, Jeremy. n.d. "Lean Artist." Accessed March 19, 2019. http://leanartist.org/.
20 "Sunderland: A City of Artists, Business and Innovation." 2017. *Creative Fuse*. March 17, 2017. http://www.creativefusene.org.uk/sunderland-city-artists-business-innovation/.
21 Graham, Beryl, Sarah Cook, and Steve Dietz. 2010. *Rethinking Curating: Art after New Media*. Leonardo. Cambridge, MA; London: MIT Press.
22 Davies, *Hackerspaces*.

Bibliography

Birtchnell, Thomas, Tillmann Böhme, and Robert Gorkin. 2017. "3D Printing and the Third Mission: The University in the Materialization of Intellectual Capital." *Technological Forecasting and Social Change*123 (October): 240–249. doi:10.1016/j.techfore.2016.03.014.

Burrough, Xtine, ed. 2012. *Net Works: Case Studies in Web Art and Design*. New York; London: Routledge, 5.

Cunningham, Caitlin. 2017. "Conflicted Commons: A Local Makerspace in the Neoliberal City." Master of Science, Richmond, VA: Virginia Commonwealth University. https://scholarscompass.vcu.edu/cgi/viewcontent.cgi?article=5874&context=etd.

Driscoll, Kevin. 2012. "The Dark Side of DIY – Makerspaces and the Long, Weird History of DIY Hobbyists & Military Funding." University of Southern California. *Civic*

Paths (blog). http://civicpaths.uscannenberg.org/the-dark-side-of-diy-makerspaces-and-the-long-weird-history-of-diy-hobbyists-military-funding/.

Hatch, Mark. 2014. *The Maker Movement Manifesto: Rules for Innovation in the New World of Crafters, Hackers, and Tinkerers.* New York: McGraw-Hill Education.

Mateos-Garcia, Juan, and Jonathan Sapsed. 2011. "The Role of Universities in Enhancing Creative Clustering." Brighton Fuse: Enhancing the Creative, Digital and Information Technology Industries (CDIT) in Brighton- an AHRC Project.

Taylor, Nick, Ursula Hurley, and Philip Connolly. 2016. "Making Community: The Wider Role of Makerspaces in Public Life." In *Proceedings of the 2016 CHI Conference on Human Factors in Computing Systems*, 1415–1425. https://chi2016.acm.org/wp/.

Umile, Dominic. n.d. "A Greater Interest in STEM Education Means a Brighter Future for Everyone." *Media Planet: Education and Career News* (blog). http://www.educationandcareernews.com/stem/a-greater-interest-in-stem-education-means-a-brighter-future-for-everyone.

+ CONTEXT I

Histories and Futures

1

ART HISTORY HACKED

Art Hack Practice as an Intra-garde

Abigail Susik

The key feature of Art History's disciplinary business is to create typologies and genealogies that strive to tell a unified story of art production as a meta-discourse across time and space, typically with a focal point located in the West. While this art historical business of creating Western-slanted, teleological narratives often makes for a persuasive hegemonic story, it easily lends itself to a facile 'grab bag' approach to history, wherein artistic precursors are continually re-cast by descendants in variable molds that serve to legitimize aesthetic practice from an institutional and imperialist point of view. Rather than setting up a one-to-one genealogy for art hack practice (AHP) across a chronological series of affinities, this chapter attempts to name a few disparate past resonances with an eye toward differentiation within comparison.

Art hack practice cannot be called yet another avant-garde due to its distinct context and content. Although there are several apparent links between aspects of 20th-century avant-garde aesthetics and art hack practice, in my mind the most interesting result of such a comparison is AHP's heightened embeddedness within modes of capitalistic production as related to wired life in the electronic or digital milieu, which pushes the avant-garde critique of capitalism to its limit. While many of the historical avant-gardes were engaged in bringing the praxis of art closer to quotidian life, as opposed to the Modernist desire for the autonomy of art, it is AHP's profound involvement in commodity production that distinguishes it from avant-garde examples and suggests new modalities for art's role and function in social life. I suggest that instead of being an avant-garde, or art that functions as a critical revelation about daily life for the masses, AHP is an intra-garde – when and only when it consists of an infiltration of daily life that attempts to disrupt consumption and production patterns from within.

If one of the precepts of an art-hacking approach is the appropriation and re-routing of both the means of production and consumption in a post-industrial capitalist sphere, art historical narratives of historical legitimization necessarily run counter to this orientation. It seems a more interesting approach, therefore, to admit the foundation of AHP in the politicized tactics of certain 20th-century avant-gardes. At the same time, one must question how this historical filiation is itself utilized, dismembered, rerouted and transformed by contemporary hacker or maker artists who employ technology as a medium and may utilize the manufacturing facility as a studio, but who, above all, seek a creative approach to deconstruction. Those ways in which the historical avant-gardes modeled art's functionality or dysfunctionality in a capitalist realm are adapted and extended by some media artists today not just as a result of their location in a different era and milieu, but also arguably because a 'hack' orientation creates separation and critical distance even with centers of affinity.

This approach of scattered comparison toward an end of differentiation is also a methodological necessity. This is so because given some of the distinctive characteristics of art hack practice, such as its aforementioned embeddedness within modes of both industrial production and consumption, there are in fact few obvious avant-garde precursors available that fully match up for a typological 'fit.' I will, however, argue in an overarching manner that AHP can be considered an extension of the avant-garde approach to culture that formulated around the end of the 19th century in part as a critical response to the early stages of mass production and information processing in industrial capitalism. Although the various avant-gardes in Europe and elsewhere around the globe answered the base/superstructure shifts in mass-industrialized society in a range of sometimes quite divergent reactions, most of them shared a core concern with the vicissitudes of capitalism in modern life and resultant aftereffects in social life. Likewise, AHP is in part a dialogic response to, an exchange with, life as it is lived now in late-capitalist nations: wired, commodity-based, instrumentalized in labor regimes that fuel production and consumption.

It is not enough to consider separate aspects of AHP as a means of setting up a comparison with past avant-gardes, because art hack practice, as I suggest above, often treats its own aesthetics as a figure for its tactics. In other words, it appears that art hack aesthetics are hacked in themselves – an appropriated bricolage of and strategies and processes applied toward a conceptual advantage. To say that technology is the baseline metaphor between art hack practice and avant-garde precursors, for example, is not sufficient because, on the one hand, technology is only one facet of the art hack approach, and in my opinion, not even the most distinctive one. Nevertheless, I will pursue here a discussion that does compare some but not all aspects of AHP to various AVGs (avant-gardes), in order to create a fragmented genealogy that simultaneously points toward AHP's hacked approach to its own historical basis, as well as the new and unique qualities of the bricolage-assemblage it makes of those historical precedents.

This approach creates a historical assessment in its attempts to situate AHP in relation to the past, since to claim that AHP is founded in AVG aesthetics and tactics but is ultimately distinct from them, and also to state that there is significant historical continuity between the early 21st century and the 12 or so previous decades – although there are also many nodes of discontinuity. For me, one essence of that discontinuity is the elaborate embeddedness of digital technology in daily life for those late-capitalist nations that are the main sites for AHP, and the intimacy human subjects share with not just technology, but also its consumption in a broader commodity network.

* * *

I will discuss five traits of contemporary AHP and make broad comparisons with AVG precedents, which will create a rapid and fragmented genealogy and also form the substrate of the differentiation from these precursors, with which I will conclude. As a historian and not a maker, my observations will no doubt be lacking in some regards. Certainly, my list of five traits is only a partial picture. The five traits of AHP I have chosen are the ones that most clearly resonate with AVG histories, which are mostly European and American, but in certain cases, such as that of surrealism, are international and transtemporal. Due to the compressed nature of this overview, my remarks are preliminary and will not provide detail about either the AVG movements I list, or individual AHP works that might resonate with them.

The five traits shared between AHP and AVGs that I will discuss are listed in order of increasing conceptual distance from AVG practice, although each of these five components has clear links to the AVG. These components overlap with one another to some degree, which makes their ordering as a list artificial and unsatisfactory. They are more appropriately dialectical nodes which interact with one another in an array of regards. The five traits of comparison between AHP and the AVG are (1) technology as content and practice/medium; (2) the appropriation and/or use of industrially produced commodities in image, material and/or process; (3) the work of art as a constructed and/or deconstructed functional or dysfunctional tool; (4) production means and/or location, often beyond the strict confines of what is usually considered the milieu of fine art (extra-aesthetic), in materials or place, and often made with shared, collective tools, knowledge or space and (5) the work of art as a commodity designed with commercial industrial tools and components, which may closely resemble a general marketplace commodity, but ultimately remains a rarified art commodity.

* * *

First to be discussed in the comparison of AHP to AVG: #1, **technology as content and practice/medium**. A technological focus is central to AHP and is likewise a key impetus and theme for AVG histories. Even so, this is a potentially misleading area which has witnessed skirmishes between historians and techno-optimists who disagree about what is new, if anything, about 'new media.' To be sure, art has been construed as *techne*, or a kind of skilled craft/knowledgeable making, since the first writings we have on aesthetics from ancient Greece. However, this art *techne* is not necessarily tied to technology as we understand it today and cannot characterize the hardware/software aspects of digital technology and the impacts that the digital has upon many areas of life. Nevertheless, the current concept of art itself is still in many ways linked to this ancient notion of a skilled making, as well as the extensions of master craftsmanship into the Renaissance, and even as far ahead in time as the cult of artistic genius which reigns for a few centuries thereafter in Europe. In addition, the Italian and Northern Renaissance are excellent examples, although earlier movements in the ancient or Medieval world would also suffice, of how art could encompass the technological in materials/process (the invention of oil paint; the development of one-point perspective as a virtual projection device etc.) as well as content or application (warfare, infrastructure, transport etc.).

I agree with many historians that with the advent of industrial production and pockets of mechanized living starting in the 18th century, as well as the range of effects this shift enacted, the definition, role and characteristics of 'fine art' began to change. In the mid- to late 19th century the notion of leading art as deskilled or counter-skilled production, among other critical stances, began to circulate and gain currency in Europe. However, it is not until machinic apparatuses and mass production begin to infiltrate daily life that 20th-century artists transform technology from a subject or technique into a medium, which entails that technology, becomes both the message and form of a work of art. Marcel Duchamp is the pivotal example for AHP, because, unlike the Italian Futurists who made technology their muse, Duchamp absorbs art into technology as medium. He depicts the machinic, he employs technology, many works are apparatuses and his general aesthetic theory is bound to the notion of a human machine. This Duchampian engine powers most of dada, which then ricochets into surrealism, where technological content is retained to some degree for critique, but as withdrawn from its identity as artistic medium. Other foci on the industrial or the thematic of the technological abound in international AVG art from the first half of the 20th century. These include, for example: Cubo-Futurism, the Bauhaus, New Objectivity, Suprematism and Constructivism, and other movements. Also relevant are so-called 'neo-dadas' such as post-World War II examples like aspects of Black Mountain College production, as well as developments such as Nouveau réalisme, Fluxus, Experiments in Art and Technology (E.A.T.), Pop Art, the genesis of video and computer art – and beyond, moving into the contemporary era. More examples of what might be called the 'technological avant-garde' can be found elsewhere in this book.

AHP is deeply indebted to this overarching trend in modern art, and especially to Marcel Duchamp and E.A.T., which added a distinct engineering element to art production. Yet already by adding this new association, that of engineering or designed functionality and application toward consumption ends, we can see that reducing AHP to #1, technology-as-medium, is insufficient for either typological or genealogical purposes. For alone in itself, technology as an artistic subject and medium does not necessarily include practical and social questions about means and ends, which concern AHP. This brings us to the second trait to be discussed: **the appropriation and/or use of industrially produced commodities in image, material and/or process**. AHP extends the AVG radical gesture of appropriating commercial culture in part or whole into the realm of art by gleaning some or all of its materials and sometimes its forms and functions from the pre-existing commercial realm. This aesthetic was initially a Realist tactic of social critique in the 19th to mid-20th century (from Courbet's political Realism, to Impressionism, Cubism and into Duchampian networks of dada and beyond), but in some cases, such as limited instances of surrealism and Pop Art, it also moved into a less-dialectical realm of the simulacral and extra-Realist (note that by Realism I am referring to the art historical use of this term as political radicalism and social themes such as labor in art, rather than the more colloquial association of 'realism' today with extreme naturalism). Already by taking technology as its medium, AHP approaches a kind of Realism through its concern with everyday life through utilized tools (implements of work and daily life). But by adding the appropriation of commercial culture into this mix, AHP strengthens its ties to the critical aesthetics of Realism by revealing itself to be mirroring certain valences of quotidian life.

These alliances are further fortified by the third trait isolated for the purpose of this comparison between AHP and AVG precursors: **the work of art as a constructed and/or deconstructed functional or dysfunctional tool**. Not all AHP works can be considered functional or dysfunctional apparatuses, but the majority of such works at least accomplish some kind of task, if only the simplest operation. AHP works are typically constructed from an array of disparate technological components, which are sometimes drawn from deconstructed elements taken from elsewhere (hence its frequent identity as bricolage-assemblage: constructed through the accretion of disparate and gleaned sources). AVG culture was avidly invested in the creation of the paradoxically dysfunctional apparatus, which most of Marcel Duchamp's readymades exemplify. This was also extended into much of dada representation and construction, such as the machine drawings of Francis Picabia or Man Ray's photographs of hybrid implements. It is also true of many later examples, such as the wonderfully defective kinetic (functionally dysfunctional) art of Jean Tinguely and others. AHP artworks, however, seem in general to be more closely allied to the idea of art as a largely functional tool that has the potential and usually does accomplish some task(s) (even if this functionality is routed toward critique of functionality itself). This would then suggest the centrality of less historically prominent, but nevertheless important, tradition of operative apparatuses in the AVG, such as Duchamp's optical works as well as those of László Moholy-Nagy and E.A.T. This point brings us already to the threshold of an attendant question that might ultimately determine the nature and extent of AHP's rapport with historical avant-gardes, or even with Modernism and Post-Modernism at large. For if AHP centers itself upon the production of tools, the question remains, what is the larger cultural purpose of these tools, functional or not, constructed or fully deconstructed, in relationship to commodity saturation and technological dependence in our lives now?

As AHP's intimate collusion in the process and results of the capitalistic production of commodities is increasingly highlighted, its proximity to AVG precedents decreases in measure. I claim that AHP's collusion in capitalistic production is anchored in the AVG critique of society through Realism, but I also think that ultimately there is a distinction that creates a rift between AHP and the past. This commensurate distance is a reflection of historical changes over the last century – human cooptation into consumption regimes in the 21st century far exceeds the beginnings of such saturation at the birth of mass production at the fin de siècle. Yet, despite this mirror effect, I am suggesting between art and zeitgeist, a catalyst theory of art as inherited from the radical streams of the AVG also impacts AHP in its embeddedness in commodity systems, as seen in work that takes place in hacker and maker spaces, as well as in personal or institutional studios or workshops. This notion of AHP as exhibiting an inherited position of agency in relation to its zeitgeist is best exemplified, for me, in the fourth trait of comparison between past precedents and our present subject: **production means and/or location, often extra-aesthetic, in materials or place, and often made with shared, collective tools, knowledge or space**. Because art hacking is not just a breaking down of existing components or commodities, but also their reconfiguration into something technologically operative (even when dysfunctional), AHP is necessarily embedded at some point of its process in systems of production as well as consumption. The artist's workshop, which was replete with different kinds of technology even in the ancient world, has more ties with pre-Enlightenment art than the AVG. Social collectives and groups rose to the fore with the earliest AVG stirrings in the 19th century – nearly all of the AVG "isms" display this collectivism. Yet, such relational grouping is not sufficient criteria for comparison with AHP, which typically entails some kind of shared space, tool or knowledge, but nevertheless may in some instances occur in solitude or physical isolation. The heart of trait #4 is that process is extra-aesthetic to some degree; process happens beyond the confines of the categorical sphere of fine art. Here again with the question of the extra-aesthetic, Marcel Duchamp's example of moving the production of art into the commercial sphere of the store with the purchased readymade is exemplary (and retains synchrony as 'hacked'), and the Bauhaus also rises to the fore with its industrial-commercial workshop model. E.A.T. and Warhol's factory also resonate, among other examples. However, AHP's investment in commercial production means and extra-aesthetic locales, along with its think-tank collective approach to knowledge or making – its manufacturing orientation – is ultimately a significant step away from AVG precedents due to the depth of this filiation with non-artistic means and process.

The fifth and final trait of comparison between the AVG and AHP is **the work of art as a commodity designed with commercial industrial tools and components, which may closely resemble a general marketplace commodity, but ultimately remains a rarified art commodity**. As aforementioned, most AVG examples engaged with capitalism in some manner, so this point is potentially misleading. The distinction is that AHP frequently creates works of art, often destined outright for an institutional context, that either resemble commercial products or closely reflect aspects of existing mass-produced products. Here, AHP's ties with surrealism return, particularly the medium known as the surrealist object (such as Meret Oppenheim's fur-covered tea cup, *Object*, of 1936) which is considered by many historians to have exerted a profound influence on advertising languages, despite surrealism's trenchant anti-capitalism. The post-World War II appropriation of commercial culture that Pop Art enacts is an under-acknowledged extension of surrealism's immanent critique of capitalism, although many surrealists condemned Pop. I argue that many, but not all, examples of art-hacked works continue the AVG critical tactic of appropriation and rerouting through AHP's entrance into capitalist production modes as well as its continual engagement with the trappings of the commercial product. At the same time, this continuation of AVG appropriation also becomes a differentiation from the AVG due to its unique level of intensity, what I have called 'embeddedness' here, with processes of commercial production. The level of appropriation of commercial means is greater in frequency and intensity in AHP than in most AVG examples, while at the same time, it is arguable that the rerouting or turning-away-from that lies at the heart of subversive appropriation as a form of societal critique, is comparably less prominent than it was with examples such as Duchamp's readymades.

AHP is therefore not an AVG, even though the relation to such a past is formative for hacked art in several regards, as we have seen. If one definition of the AVG was to be 'avant' or ahead of the game, AHP strikes me as having the potential to be an intra-garde, something wedged right into the middle of things – an infiltration into capitalist production and consumption patterns – in the trenches of the battle between subjective agency and capitalism, rather than on the frontlines. Such an entrenched approach to cultural critique has its own challenges and advantages. An intra-garde disruption from within is a covert tactic of camouflage and integration, potentially taking place over time, rather than a brash and violent attack through sudden confrontation. Rather than just the operational spin of aesthetic appropriation, which is typically a re-routing of function or outcome, intra-garde tactics involve systemic and gradual sabotage. Not all AHP is intra-garde, but there is great potential and even greater need for such a critical approach by works of art that undertake dialog with the digital.

2

THE AFRIKAN MAKER

Hacking Our Way into a Hybrid Future...

Mugendi K. M'Rithaa

Afrika's Rich Material Culture Heritage

Afrika has been home to myriad craft and material culture traditions. The so-called ethnic carvings made of wood, terracotta, bronze and soapstone are well known.[1] Whereas wood was used to make stools, statuettes and masks right across the continent, certain communities such as the Kamba (of Kenya) and Makonde (of Mozambique and Tanzania) are better known for their exquisite woodcarvings and sculpture made of indigenous hard woods that are durable and rich in symbolism.[2] The Kongo, Luba (of the Democratic Republic of Congo) and the Loango (of Angola) were also excellent craftsmen who excelled in a number of natural materials such as wood and ivory.[3]

The Yoruba of Nigeria are renowned for their terracotta and bronze busts.[4] Additionally, the Shona of Zimbabwe and the Kisii of Kenya make beautiful artefacts out of local soapstone varieties.[5] The Zulu, Ndebele and Xhosa of South Africa are renowned for their colorful beadwork, as are the Maasai of Kenya and Tanzania.[6] Further, the Baganda of Uganda produce various items from the indigenous eco-friendly barkcloth from the Mutuba tree (*Ficus natalensis*);[7] while the self-same Baganda and Ashanti (of Ghana) are renowned for their colorful printed fabrics, with the later making the world famous woven *kente* cloth.[8]

Whereas one cannot speak of a pure Afrikan aesthetic, Stéphane Guibourgé,[9] however, identifies some key overarching themes and elements that are consistent across the continent, *inter alia*: hybridity; informality; hackability; reusability; natural materials; earthy colors; geometric patterns as well as organic shapes and forms. Tapiwa Matshinde[10] challenges the narrow perceptions on aesthetic expressions from the continent by showcasing this richness. To this end, the

Design Indaba[11] (the Southern Hemisphere and Afrika's premier annual design conference hosted in Cape Town since 1995) focuses on the continental and global best practices in 'creativity, through the lens of the work and ideas of leading thinkers and doers, opinion formers, trendsetters and industry experts' by showcasing contemporary exemplars of 'design for the 99% – with particular emphasis on creativity from Africa – and design that's linked to improving the quality of life.' Similarly, the definition of Industrial Design advanced by the WDO[12] focuses on an arguably more holistic mission of helping to improve (or better) the 'quality of life' (as opposed to that of merely raising the standard of living).

The Maker Movement in Afrika

Afrika is home to many informal manufacturing sectors such as the vibrant ones in cities like Lagos, Accra, Harare, Kampala and Nairobi (where it is known as the *jua kali* sector[13]). The Maker Movement in Africa arguably has its origins in the introduction of Fablabs – fabrication laboratories that were introduced in various cities across the continent as championed by Neil Gershenfeld[14] with a vision of catalyzing and democratizing personal fabrication to unlock the continent's vast potential for innovation.[15] The Fablabs ushered in a peer-to-peer ethos that is particularly germane for the African context. The typical open-source rapid prototyping equipment found in such Fablab workstations includes *inter alia*: 3D printers, laser cutters, milling machines, wood and metal lathes, band-saws, vinyl cutters, embroidery machines as well as basic metal-cutting and welding equipment.[16]

The Maker Faire Africa (MFA)[17] events were first hosted in Accra (MFA 2009) to showcase ingenuity and creativity on the continent. Subsequent events have been celebrated in Nairobi (MFA 2010); in Cairo (MFA 2011); in Lagos (MFA 2012); in Johannesburg (MFA 2014); in Cape Town – including a conference (MFA 2015); as well as major international events in 2013 in Istanbul, Milan and New York. The (second, third and tenth) goals of the MFA manifesto boldly states that 'we will make the things Africa needs'; 'we will see challenges as opportunities to invent, and invention as a means to proving African ingenuity'; and 'we will remake Africa with our own hands.'[18] The MFA initially set out to develop an ecosystem on the continent – the change observed over the last decade has resulted in a change of focus as continent-wide events are no longer deemed necessary due to the wide diffusion of the movement.

The Maker Station in Cape Town

Felix Holm[19] was born in Pretoria 'into a family of artist, artisans, designers, musicians and craft producers, generations deep in all directions of the tree.' He joined the Maker Movement in South Africa in 2013 and opened the Maker Station[20] in Cape Town with his brother the following year. Holm[21] studied Industrial Design at the Cape Peninsula University of Technology (CPUT) before working as a prototyping, furniture design and in the exhibition industry – he has experience working with local and international donor organizations operating in deep rural communities to set up small-scale manufacturing, craft and distribution businesses.

Cape Town's status as a design-friendly city was significantly elevated when it was awarded the prestigious designation of World Design Capital™ (WDC) 2014 by the World Design Organization (WDO).[22] Cape Town is the first African city to be designated as both a WDC in 2014, and as a City of Design as recognized by the United Nations Educational, Scientific and Cultural Organization (UNESCO) Creative Cities Network[23] in 2017. Cape Town is home to a diverse range of creative industries mainly operating within the vibrant Woodstock and Harrington Street areas – the Maker Station is located in the former.

The Maker Station acts as a catalyst within the craft, design and fabrication ecosystem of Cape Town as well as coordinating the local activities with those happening elsewhere within the country and region. Holm concurs with the author on the belief that the 21st century is Afrika's to claim. According to Holm,[24] the Maker Movement can support this vision by offering "various 'tools' by virtue of its core tenets of sharing and creating access, to education firstly and secondly as an enabler for conventional small business and thirdly for innovation in a more tech leading edge space." Holm[25] further views the key elements that distinguish the Maker Movement in Afrika from those in Europe or elsewhere as the following:

- The huge diversity in the demographic of users, economic status, education levels, access to transport and related amenities;
- Maker spaces in the industrially developed parts of the world have a much more homogenous audience – this changes the business model and offering needed so fundamentally that it almost unrecognizable elsewhere;
- In industrially developing (or majority world) contexts (where 90% of humanity subsists), the vast majority of makers enter the trades as a subsistence or survival strategy, not with high-end innovation as first priority where in the [developed economies] most makers are middle-class, with better education, higher disposable incomes and easier access to other financial resources (such as seed capital for design and development) among other advantages and
- The emphasis in majority world contexts is on access to basic technological tools, whilst in developed settings, the greatest value for makers working in communities is on the rich inter- and poly-disciplinary mix of talent at their disposal to foster creative collaboration.

A number of unique products and technological trends in have been identified within the Cape Town ecosystem. These include a number of technology start-ups mainly within the vicinity of the Maker Station in the Woodstock area such as HealthQ, Lukami, Nomanini and some other smaller projects. Such development projects typically exhibit a DIY orientation to innovation wherein makers 'make a plan with limited available resource.'[26] The sharing of technical skills and resources in this informal business development environment is well established within the local Maker community.

Enriching the Ecosystem

According to Suné Stassen[27] the Co-Founder, Festival and Program Director of *Open Design Afrika*,[28] the traditional view of *STEM* (Science, Technology, Engineering and Mathematics) has limited efficacy as it omits the critical skills of creativity. Stassen[29] advocates a *STEAM* approach (that includes 'A' for Art and Design). As Stassen[30] further argues:

> Art and Design are poised to transform economies and push innovation in the 21st Century. Collaboration and exercising a holistic outlook will guide us towards a more sustainable and effective answer to the question: "What skillset is required for the 21st Century?" What we know for sure is that problem-solving, dynamic, creative and innovative process thinking, will remain at the forefront of the required skills in the 21st Century. [...] Creative skills have finally been recognized as the core ingredient to add value and meaning to business, products, systems, environments and even service-oriented markets. It is clear that integration between different subjects and study fields and cooperative learning have become vital elements to drive a shift in education from STEM to STEAM.

Holm[31] concurs with this view as such an inspired approach would 'promote sound business practice guidelines' seen to be amenable to the development and enrichment of the Maker Movement. Additionally, both Holm[32] and Stassen[33] argue for a more dynamic Quadruple Helix partnerships (of academia, business, government and civil society) to ensure the sustainability of the Maker sector, preferably with formal links to the broader creative and manufacturing industries. To this end, it is worth noting that the Maker Station in Cape Town does indeed engender this rich Quadruple Helix composition in its active strategic partnerships.

Further, the Maker Station acknowledges the goodwill it has enjoyed with various partners, including the donations of key equipment from the Vaal University of Technology (VUT) and the Central University of Technology (CUT). The mentorship provided by respected academics within the advanced manufacturing and 3D printing fields has benefited the station significantly.

The regular visits and participation by international experts such as Jesper Kildegaard Jacobsen[34] from Denmark further enrich the human and technical resource capacity of the Maker Station. Jacobsen[35] is a multidisciplinary designer, innovator and entrepreneur and is an experienced Fablab manager with work experience in Kenya, Rwanda, Uganda and South Africa. Jacobsen[36] has engaged with the Maker Movement via the Republikken[37] (The Republic) Maker space, as well as the highly successful *Fablab Nordvest*[38] in Copenhagen. He is currently leading a fundraising initiative by the Fablab Nordvest management to buy a robot arm 'which is right on the peak of new technology in the world today. That will give us a big leap in front of the other Fablabs/Maker spaces not only in Copenhagen but also Denmark in general.'[39]

Jacobsen[40] views the Maker Movement 'as one of the biggest contributions for the changes to come' and hopes 'the changes needed, will be implemented in a progressive and beneficial manner for the future population in Africa.' For the continent to achieve its developmental potential, Jacobsen[41] believes that the strategy for the Maker Movement in Africa should be carefully considered, well planned and professionally implemented in the various countries so as 'to suit the culture, traditions and skills of the specific area,' and proposes that the Maker Movement helps educate, innovate, problem-solve and develop by doing the following:

1. New product;
2. Production facilities and production processes;
3. Process analytics and optimization;
4. Product development and innovation;
5. Worldwide networking;
6. Management and
7. Education.

Conclusion: Made in Afrika

As Chris Anderson[42] argues, 'any country, if it wants to stay strong, must have a manufacturing base.' This is as true for a country as it is for a region (or continent in our case) to ensure economic and technological growth and sustainability. To achieve this noble end though, the means must be in alignment. The resources required are of the following nature:

- *Education, technical skills and vocational training*: the educational offerings (from as early on as primary/elementary school) should incorporate the full ambit of STEAM subjects so as to ensure creative and innovative problem-solving. In addition, a STEAM approach facilitates transdisciplinarity within pedagogical, technical and professional domains;

- *Expos, fairs and public engagement*: accessible expos, festivals and competitions could complement existing strategies for the diffusion and dissemination of creative outputs by generating greater public awareness of the benefits of a Maker culture;
- *Planning, policy and governance*: Governments from across the continent should be ambitious, yet realistic in the framing of the developmental plans by prioritizing manufacturing, maths, literacy and entrepreneurial strategies for economic growth;
- *Quadruple Helix Partnerships*: the various arms of government (from local to national) should embrace the catalytic potential of tapping into the business/industrial sector, civil society, and academia in a holistic and integrated manner as engines for long-term growth; and
- *International collaboration*: local Maker communities should engage more robustly with the global Maker ecosystem movement to learn from the latter's best practices, latest technological advances as well as to generate an adaptable range of context-responsive tools and solutions for wider diffusion within an open-source peer-to-peer ethos. This could be achieved via a unified, dynamic and accessible portal for all participants.

The training of future designers, crafters, makers and artisans will need to anticipate current developments wherein the role of the designer has fundamentally changed.[43] As a region in transition, investments in new tools, technologies and capabilities should be implemented, complimented by better education, access to information and learning via the *Internet of Things*.[44] Such proactive strategies would ensure Afrika's robust participation in the *Fourth Industrial Revolution* – not merely as followers, but potentially leapfrog within specific aspects of making.[45]

Leapfrogging is not merely a pipe dream, as recent developments have demonstrated. Afrikan design innovations are indeed making a difference as evidenced through leapfrog solutions that include *inter alia*: the world's first drone-port for delivering of medical supplies in Kigali; the *Enda* athletics running shoes created by marathon runners in Kenya; the *MPesa* mobile money payment system developed in Nairobi; *Kiira* fully electric car and bus in Kampala; animators like Wesley Kirinya of *Leti Arts* (Accra and Nairobi); furniture designers like Peter Mabeo (Gaborone), Bibi Seck (Dakar) and Haldane Martin (Cape Town); architects like Kunlé Adeyemi who designed the ingenious Makoko floating school (Lagos); experimental robot makers like Ralph Borland (Cape Town); fashion designers like Laduma Ngxokolo who designed the culturally inspired Maxhosa knitwear range (Cape Town and Johannesburg); and socially conscious industrial designers like Byron Qually (Cape Town).

The aforementioned trends bode well for a continent rich in natural talent, raw materials and large quantities of other valuable resources that are ideal for supporting the Maker Movement in Afrika. Such factors imply means that increasing numbers of people on the continent have the potential to co-produce novel and innovative inventions and technologies that could find adoption further afield. The Maker Movement in Afrika has the potential to foster a paradigm shift wherein the next generation of makers adopt a decidedly *prosumer* ethos with a marked and resilient entrepreneurial bent.

Notes

1 Guibourgé, S. 2000. *African Style*. Paris: Flammarion.
2 von D. Miller, J. 1975. *Art in East Africa*. London: Fredrick Muller Ltd.
3 http://www.contemporary-african-art.com/african-sculpture.html
4 https://za.pinterest.com/pin/486177722256349453/?lp=true
5 http://www.zawadee.com/blog/ca/soapstone-carvings-by-the-kisii-people-of-kenya
6 https://za.pinterest.com/kerneels59/african-beadwork/?lp=true
7 https://ich.unesco.org/en/rl/barkcloth-making-in-uganda-00139
8 https://kwekudee-tripdownmemorylane.blogspot.co.za/2012/12/kente-cloth-ghanas-ashanti-cultural.html
9 Guibourgé, *African Style*.
10 Matshinde, T. 2015. *Contemporary Design Africa*. London: Thames & Hudson.
11 http://www.designindaba.com/about-design-indaba
12 WDO. 2015. *Definition of Industrial Design*. World Design Organization. http://wdo.org/about/definition/ [Accessed: February 20, 2018].
13 *jua kali* literally means "in the hot sun" in KiSwahili – a language widely spoken in East and Central Africa.
14 Gershenfeld, Neil. 2005. *FAB: The Coming Revolution on Your Desktop- From Personal Computers to Personal Fabrication*. New York: Basic Books.
15 M'Rithaa, M.K., Verveckken, B. & Futerman, R. 2008. Shifting trajectories: advancing cosmopolitan localism through participatory design. pp. 1145–1154 in: C. Cipolla and P.P. Peruccio (eds), *Proceedings of Changing the Change: Design, Visions, Proposals and Tools*. Alemandi Conference Press. July 10–12, 2008. Torino. Available at http://www.allemandi.com/university/ctc.pdf [Accessed: January 28, 2018].
16 Ibid., Holm, F. 2018. Electronic responses to semi-structured interview questions from the author. Cape Town. February 14, 2018. [Email correspondence in possession of author].
17 http://makerfaireafrica.com/about/
18 http://makerfaireafrica.com/maker-manifesto/
19 Ibid.
20 http://makerstation.co.za/
21 Ibid.
22 http://wdo.org/programmes/wdc/past-cities/wdccapetown2014/
23 https://en.unesco.org/creative-cities/events/64-cities-join-unesco-creative-cities-network
24 Ibid.
25 Ibid.
26 Ibid.
27 Stassen, S. 2018. Electronic responses to semi-structured interview questions from the author. Cape Town. February 13, 2018. [Email correspondence in possession of author].
28 http://opendesignafrika.org/
29 Ibid.
30 Ibid.

31 Holm, *Electronic responses.*
32 Ibid.
33 Stassen, *Electronic responses.*
34 Jacobsen, J.K. 2018. Electronic responses to semi-structured interview questions from the author. Cape Town. February 19, 2018. [Email correspondence in possession of author].
35 Ibid.
36 Ibid.
37 https://republikken.net/
38 http://fablabnordvest.dk/
39 Ibid.
40 Ibid.
41 Ibid.
42 Anderson, C. 2012. *Makers: The New Industrial Revolution.* London: Random House Business Books: 23.
43 Manzini, E. 2015. *Design, When Everybody Designs: An Introduction to Design for Social Innovation.* Boston: MIT Press.
44 Jacobsen, *Electronic responses.*
45 Holm, *Electronic responses.*

Bibliography

Boon, M. 2007. *The African Way: The Power of Interactive Leadership.* Cape Town: Zebra Press.

Cape Town NPC. 2015. *World Design Capital Cape Town 2014: The Story of an African City.* Cape Town: Paul Duncan Media (Pty).

de Beer, J., Armstrong, C., Ellis, M. and Kraemer-Mbula, E. 2017. *A Scan of South Africa's Maker Movement.* Working Paper 9. Open African Innovation Research (Open AIR) network. Available at: http://www.openair.org.za/publications/a-scan-of-south-africas-maker-movement/ [Accessed: February 18, 2018].

Eggers, W.D. & MacMillan, P. 2013. *The Solution Revolution: How Business, Government, and Social Enterprises are Teaming Up to Solve Society's Toughest Problems.* Boston: Harvard Business Review Press.

Gade, C.B.N. 2017. *A Discourse on African Philosophy: A New Perspective on Ubuntu and Transitional Justice in South Africa.* Lanham: Lexington Books.

Kries, M. & Klein, A. (eds). 2015. *Making Africa: A Continent of Contemporary Design.* Weil am Rhein: Vitra Design Museum GmbH.

Mafundikwa, S. 2006. *Afrikan Alphabets: The Story of Writing in Afrika.* Brooklyn: Mark Batty Publisher.

Manzini, E. & M'Rithaa, M.K. 2017. Distributed systems and cosmopolitan localism: an emerging design scenario design scenario for resilient societies. pp. 29–37 in: A. Skjerven and J.B. Reitan (eds), *Design for a Sustainable Culture: Perspectives, Practices and Education.* London: Earthscan (Routledge).

M'Rithaa, M.K. & Jamie, A. 2017. Advancing the Afrikan lions' narrative: the quest for a sustainable future for all… *Proceedings of the 24th Annual FIDIC-GAMA Conference.* 7–10 May 2017. Cape Town. Available at: http://www.cesa.co.za/sites/default/files/GAMA2017_E3_Advancing%20the%20Afrikan%20lions%20narrative%20final_Prof%20Mugendi.pdf [Accessed: January 20, 2018].

Wanda, R.E. 2013. Afrikology and Community: restorative cultural practices in East Africa. *The Journal of Pan African Studies,* 2013 (6):1–25.

3

REPROGRAMMED ART, A BRIDGE BETWEEN THE HISTORY OF INTERACTIVE ART AND MAKER CULTURE

Serena Cangiano, Davide Fornari and Azalea Seratoni

Introduction

In September 2014, five artists and designers met in Lugano, Switzerland, to conduct an experimental journey through time that connects Gruppo T's kinetic and Programmed Art, a crucial episode in the history of the arts, with contemporary practices linked to maker culture and open design. The experiment took place within *Re-programmed Art: an Open Manifesto*, a project that examined two worlds chronologically distant from one another by 50 years. In doing so, it served to update the discourse on the relation between early interactive art and the opportunity to make an art for everyone that can be replicated freely and collaboratively.

This research sets out to report on this experiment, which aimed to hack historical artworks. These are reinterpreted and redesigned according to the tenets and practices of the maker and open source movements. Specifically, this chapter presents the principles of the pioneering artistic production of Gruppo T, a collective of artists working at the end of the 1950s who defined the basis of what we today call interactive art. Hacking the works by Gruppo T in the framework of an action-research project allows us to reconsider open source practices within artistic contexts by comparing them to Gruppo T's experiments. The presentation of several prototyped artefacts concludes this chapter, together with a reflection on prototyping practices within the maker community in a context where the impact of technologies on society is considered through the arts.

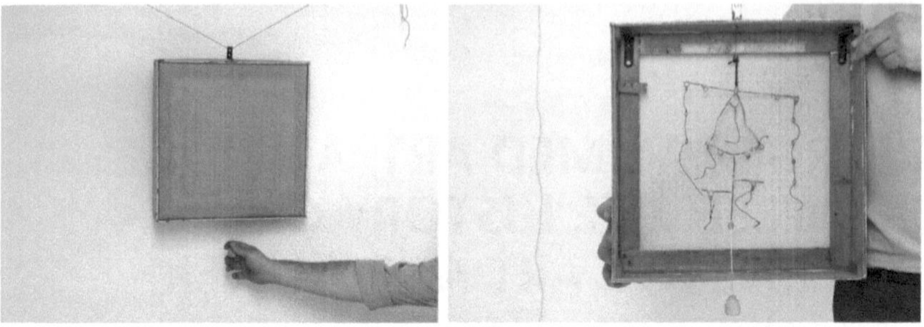

FIGURE 3.1 One of the first and interactive version of Gabriele Devecchi, *Superficie in vibrazione*, [left] front [right] rear. Photo: reprogrammed-art.cc, 1959.

'Do it yourself!' Gruppo T and the Anticipation of DIY and Open Source Art

A visitor to the fairs of today's inventors, known as Maker Faires, or the spaces of collaborative and digital production, known as fablabs, may become familiar with the value of sharing a source file, the use of free licenses such as Creative Commons or the importance of process documentation in the making of a technological device. If one works today in the areas of programming, generative design, interactive artefacts and environments or with open source hardware such as Arduino, they might well stumble upon a link between the novel practices of tinkering and technological hacking, or 'Do It Together,' as proposed by today's Maker Movement, and experiments with interactive arts that emerged at the end of the 1950s.

This was in fact our own starting point to this research. We sensed that the impact of disruptive practices typical to maker culture and the open source movement could be interpreted from the viewpoint of artistic production from the 1950s to 1960s. This could be based on algorithmic approaches that were employed by artists and on the use of computers as new tools for design. As teachers, researchers and curators operating in the fields of arts and interaction design, we followed this initial intuition, starting in 2013 as an action-research project connecting kinetic and Programmed Art – one of the last taboos of art history[1] – with the living practices of maker spaces, fablabs, Maker Faires and free software and hardware.

'Programmed Art' is the definition given to a body of works by a group of Italian artists active between the end of the 1950s and the early 1960s. This definition was coined by Bruno Munari and Umberto Eco in December 1961 and used in *Almanacco Letterario Bompiani*[2] and on the occasion of the exhibition held at the Olivetti showroom in Milan in 1962, featuring works by Munari, Enzo Mari and the artists of Gruppo T (Giovanni Anceschi,

Davide Boriani, Gianni Colombo, Gabriele Devecchi and Grazia Varisco) and of Gruppo N.[3]

Gruppo T was a collective of young artists who envisioned processes similar to those of today's technological and design research methods. They did so by creating prototypes of works that were then reproduced in series as ever-changing variations of artefacts. The group rebelled against the art system by eschewing brushes and canvases in favor of works that embodied electromechanical parts and included industrial materials. Grazia Varisco, the only female member of the group, built her *Schemi luminosi* ('Light Screens') by overlapping and layering Plexiglas sheets to create ever-changing patterns. Devecchi employed iron pins moving over an elastic surface made of natural rubber, triggered by the motorized movement of a propeller on the back that served to randomize the neat alignment of the pins. Giovanni Anceschi built virtual volumes by painting black and white stripes on rods that would rotate at a fast pace. Davide Boriani, in his most well-known series of works *Superficie magnetica* ('Magnetic Surface'), played with the magnetism of iron powder to generate never-ending images. Gianni Colombo's *Strutturazione pulsante* ('Pulsating Structuration') features an ever-changing configuration of slowly moving Styrofoam bricks.

The five artists operated in industrial workshops, where they were able to access and experiment with materials. It was the beginning of the 1960s and the workshop was considered a kind of ideal artist studio, where instead of brushes and canvases, members of the group could find tools, machines and instruments fit for producing and realizing their vision of a new way to make art. As Gabriele Devecchi wrote, the members of the group were a metaphor for teamwork: artists collaborate in a procedural way in order to forge 'objects that can be reproduced by anyone, anytime.'[4] Thus, the workshops themselves were the symbolic place of teamwork for the group.

Gruppo T's collaboration brought an extreme example to the debate of authorship occurring in the art world; it suggested practices that were similar to what we now see in the ubiquitous online sharing, remixing and copying of the Internet era. Gruppo T members often co-signed their works. This enacted a collaborative process that was centered on the kinetic and perceptual effects of ideas rather than on the physical and sculptural qualities of works of art. Gruppo T's focus was on the body of the viewer, the intangible effect of an artwork on one's perception, the research process, interaction and the rules that defined and produced an iterative and cyclical realization of their works.[5]

The physical object, the kinetic sculpture, was comprised of a set of principles of knowledge. It was the result of a prototype, an experiment of artistic research. The prototype was considered open and could be replicated by anyone. This method of working is exemplified in an anecdote about Gruppo T's practice. 0 ⇔ 220 volts by Colombo is a work of extraordinary simplicity: two light bulbs enact a maximum or a minimum brightness in a continuous and slow alternation of intensity. When Anceschi saw the work for the first time, he immediately fell in love with it and asked Colombo if he could have a copy of it, to which the latter replied: 'do it yourself!' This option of reproducibility made the work of Gruppo T a scandal. It also triggered our intuition, as researchers, to examine the connection between these masters' works and contemporary artistic sharing practices.[6]

An Open Source Process to Re-enact Artworks by Gruppo T

More than 50 years after the establishment of Gruppo T, Giovanni Anceschi, a member of the group, stepped into Lugano's fablab and remarked how it looked similar to him as a wonderful toy room looks to a child:

> Compared to the poverty of means that were available to us in the 1960s, the range of materials that are at hand today is never-ending: transparent and opaque, colored and metallic, hard and soft, stiff and flexible, heavy and light, static and metamorphic, passive and even reactive.[7]

Another Gruppo T artist, Grazia Varisco, moved around Lugano's fablab, inspired by the complex shapes that can be modeled through 3D printing.

In September 2014, both artists joined a group of five young designers, artists and makers, to work together in the fablab in a way that was vastly different from their original experimental workshops in the 1960s. In the fablab, the two members of Gruppo T participated in a collaborative process that would produce five works. These were inspired by the principles of kinetic and Programmed Art, yet produced with the tools and approaches of DIY and open source typical to the contemporary Maker Movement. Together with Thibault Brevet, Fabio Franchino, Martin Froelich, Giorgio Olivero and Yvonne Weber, they participated in the re-enactment and hacking of the works of Gruppo T. In doing so, they aimed to build prototypes that would translate the tenets of Programmed Art into codes of 21st-century culture through open source hardware, software and digital fabrication technologies.

Research through making allowed the team to go beyond the production of replicas and to reactivate and reenact the subversive practice of Gruppo T. It did so by reprogramming their practices with new tools, techniques and maker processes. Each young artist/designer worked on the development of a project aimed at translating an artwork by Gruppo T, embedding this translation into new physical hardware and ensuring that suitable instructions were made available online in order to make the work replicable. Thus, others can reproduce the project, repair it or subvert it completely. The replicability of the process is ensured by the use of open source licenses, open hardware and Creative Commons licenses. This process allows for sharing and distributing information on how to create visual effects, visualize physical phenomena and interactions, manipulate gears and play with materials and technology. These are the same intangible qualities embodied by the art of Gruppo T.

Among the outcomes of this action-research, two projects exemplify this operation of translating Programmed Art within the context of maker culture. They do so by proposing 'enabling artworks.' These are machines that support people in participating in the creative process; they allow others to produce new artefacts through open mechanisms and codes.

Magnetic Drawbot by Giorgio Olivero and Fabio Franchino is a machine that generates drawings in a procedural way according to algorithmic instructions and through mechanical operations. The machine employs ferrofluid suspension as ink and a magnet as a cursor. This cursor is triggered by a microcontroller connected to a system of motors and linear guides. Conceived as a 'simple

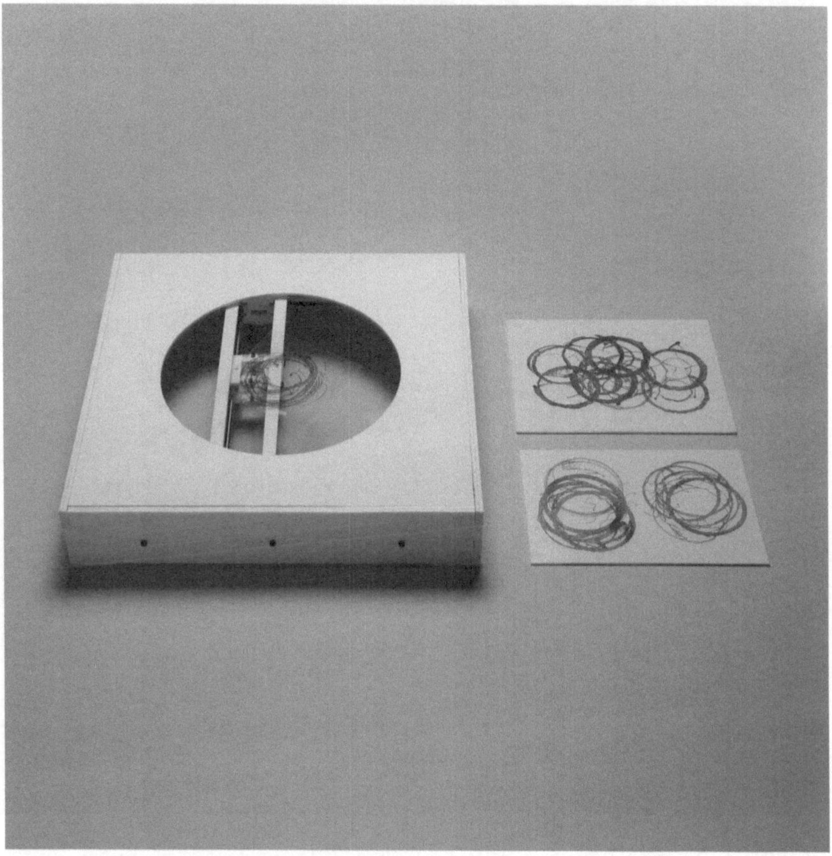

FIGURE 3.2 Fabio Franchino and Giorgio Olivero, *Magnetic Drawbot*, 2015. Reprogrammed open source derivative Davide Boriani, *Magnetic surface n. 6*, 1959. Photo: reprogrammed-art.cc.

machine,' it was low cost and easy to build, replicate, modify and adapt. *Magnetic Drawbot* is fit for reprogramming and can be adopted for educational purposes. It can become a useful example for teaching how to build other machines that draw patterns which vary over time. The machine was built by applying open source methods and 'modifying' principles.[8] It was derived from an open source 2D plotter that was altered to generate a new application.[9] A second principle applied in the project is the empowerment of the user: *Magnetic Drawbot* enables users to create their own generative art, to familiarize themselves with technology and to decode and recode the algorithm that randomizes the visual output. This reflects Gruppo T's artistic method to be open to the participation of the audience, thus introducing a novel and crucial feature that would become interactive art.

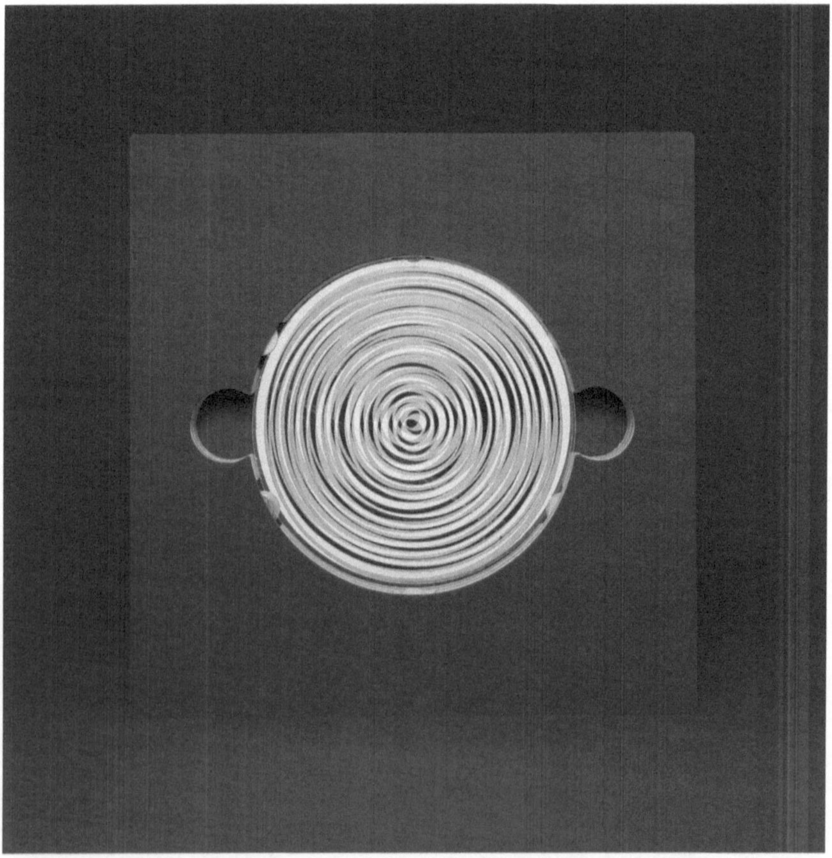

FIGURE 3.3 Yvonne Weber, *Topologia della luce*, 2015, reprogrammed open source derivative of Grazia Varisco, *Variable light screen R.R. 66*, 1962. Photo: reprogrammed-art.cc.

A second project is *Topografia della luce* ('Topography of Light') by Yvonne Weber. The artist worked on the reinterpretation of *Schemi luminosi* ('Light Screens') by Grazia Varisco, using an LCD screen as a raw material. The screen was deconstructed by removing the polarizing filter, which was reconfigured in order to create a pattern on a Plexiglas disc that users can remove and modify in order to generate custom light effects. Two motors allow the rotation of the Plexiglas disc and a Processing sketch modifies the light source of the hacked LCD screen. Each user can design a disc; this model is the basis for motivating interaction and play with the work.

Similar to *Magnetic Drawbot*, *Topografia della luce* is also a reinterpretation using maker strategies: each disc is a customizable element with a pattern laser-cut out of a polarizing filter. Digital fabrication technology, in this project, is the means to facilitate the reuse and recycling of technological waste, which hints at Gruppo T's research into new industrial materials during the economic boom of the 1960s.

Multiplied Art, Prototypes and Open Derivatives

Reprogrammed Art is an open manifesto and a platform that asks museums, cultural institutions, artists, makers and the public at large to reconsider the work of art in the age of open source reproduction. The project offers a digital repository (www.reprogrammed-art.cc) that collects and documents the work of Gruppo T. This repository facilitates the reproducibility of such a unique experience in art history, the bringing together of Gruppo T with young artists through maker tools. This becomes a useful example of modalities and methodologies for the creation and distribution of an art for all. It also hints at the period of 'multiplied art' when the market offered limited edition art multiples. This concept of multiples as art objects for large publics was first realized by an emblematic Italian movement that joined art, design and industry. Danese edizioni, a design brand established by Bruno Danese and Jaqueline Vodoz, became a pioneering workshop for the production of art multiples by Bruno Munari, Enzo Mari, Daniel Spoerri's Edition MAT and the so-called 'last vanguard' represented by the kinetic and Programmed Art movements, to which Gruppo T belonged.[10] In 2005, Gruppo T's multiples for Danese were reissued by the Italian design brand Alessi, which pushed further the utopic vision of making art accessible to all through serial, semi-industrial production.

From an open source viewpoint, the experience of multiples is the element that connects Gruppo T and our reinterpretation of the artworks. Victor Vasarely, writing about multiplied art, anticipated concepts that would sound familiar in a maker space today:

> If yesterday, art meant feeling and making, today it may mean conceiving of something and having it made. If in the past the duration of an artwork

> was based on the optimal quality of its materials, on technical perfection and manual ability, today it rests on the awareness of a possibility of replicating, multiplying and diffusing. Thus, the myth of the unique piece will vanish with craftsmanship and works that can be disseminated thanks to mechanization will triumph. We shall not be afraid of the means that new techniques have given us; we can only live in our own age.[11]

On the other hand, the Italian designer and theorist Bruno Munari wrote that multiplied art introduced new production processes based on prototyping as it occurs in the fablab:

> In multiplied art, the accessibility of art is provided by its replicability [...]. The art multiple is designed using methods of experimental design, and the person in charge of its production is not dealing with an artwork, but with a prototype that allows for serial production.[12]

Reprogrammed Art is grounded in a similar discourse: it proposes the concept of reproducibility extended through an open source approach. Along with historical documentation, the web repository created for this project offers tutorials and blueprints that facilitate the production of reprogrammed works in any digital fabrication workshop. These instructions allow a subsequent artist/maker to follow the prototyping process of the reprogrammed works. They also enable hardware and software derivatives to be created or the kinetic effects of the original piece to be activated.

Conclusions

In recent years, the ability to share knowledge over the Internet has brought a paradigm shift to manufacturing. Access to the means of digital production by publics and the establishment of interdisciplinary spaces attended by engineers, artists, designers, students and teachers have promoted a return to practical knowledge. Technology is largely widespread, and bottom-up movements such as open hardware, open software and DIY communities aim to make it accessible by enabling documentation and sharing practices.

By applying collaborative processes in the 'Reprogrammed Art' project, we learned that shared manufacturing practices present in maker spaces and fablabs allow for a return to collaborative art. As libraries of peer production,[13] they make possible peer-to-peer art that is open to a global community. This is facilitated by online access to technical standards and mostly by the adoption of open source licenses. This action-research allowed us to reactivate artistic experiences from the 1960s through the re-enactment of a collaborative process in a physical space. Applying an open source framework was essential to the project – without it, we could never have created continuity with the intentions of Gruppo T: a democratic art for everyone.

Open licenses such as Creative Commons have offered a legal context fit for the community of stakeholders interested in kinetic and Programmed Art: they allow them to produce derivative works instead of mere replicas. Derivative works bear in their code the DNA of the original work. This can then be connected to the authors of the reinterpretation and to tenets typical of the artistic production of Gruppo T. Without the use of such licenses, we would have simply hacked artworks. Instead, artworks were created that hack the art system in a disruptive way by proposing a distributed, digitized and participative method of making art. Through this operation of opening Gruppo T's work, we transformed the utopic ideas of artistic vanguards into practices of preservation, communication and hacking while defining trajectories of artistic futures.

Notes

1 Gioni, Massimiliano. 2012. "Civiltà Delle Macchine." *Domus*, no. 957 (April): 118–121.
2 Morando, Sergio (ed.). 1962. *Almanacco Letterario Bompiani 1962*. Milano: Bompiani: 3.
3 Vergine, Lea. 1983. *Arte Programmata e Cinetica 1953/1963. L'ultima Avanguardia*. Milano: Mazzotta.
4 Devecchi, Gabriele. 1983. "A Proposito Delle Ipotesi Miriorama." In *Arte Programmata e Cinetica 1953/1963. L'ultima Avanguardia*, 168. Milano: Mazzotta.
5 Seratoni, Azalea, and Davide Fornari. 2015. "Time Has Caught Up with Gruppo T." In Serena Cangiano, Davide Fornari, Azalea Seratoni (eds), *Re-programmed Art: an Open Manifesto*, 29–43. Monza: Johan & Levi.
6 Cangiano, Serena, and Davide Fornari. 2015. "Re-programmed Art: an Open Project." In Serena Cangiano, Davide Fornari, Azalea Seratoni (eds), *Re-programmed Art: an Open Manifesto*, 9–15. Monza: Johan & Levi.
7 Giovanni Anceschi in conversation with the present authors, Lugano, September 1–7, 2014.
8 Modifying a source file of a project is one of the conditions permitted by open source licenses such as Creative Commons or The MIT License. This permission allows, for example, to adapt the source code of a project to implement a new feature or to use it in another project. It is one of the permissions indicated by the definitions of open source software, hardware and design. The definition of Open Design states: "Open Design is a design artifact project whose source documentation is made publicly available so that anyone can study, modify, distribute, make, prototype and sell the artifact based on that design." The definition is available on GitHub at https://github.com/OpenDesign-WorkingGroup/Open-Design-Definition/blob/master/open.design_definition/open.design.definition.md [Accessed January 12, 2018].
9 HBot is a movement system that employs a single transmission belt and allows motors to be positioned without moving them on axes. It is used in the project 'Open Stage' by Gianluca Pugliese available at: http://www.fabacademy.org/archives/2015/eu/students/pugliese.gianluca/project08.html [Accessed January 12, 2018].
10 Vergine, *Arte programmata*.
11 Vasarely, Victor. 1960. "Edition MAT. Opera d'arte Animate e Moltiplicate." Exhibition catalogue, Galleria Bruno Danese, Milan, February 6, 1960–March 31, 1960.
12 Munari, Bruno.1971. *Codice ovvio*. Turin: Einaudi: 52.
13 Troxler, Peter. 2011. "Libraries of the Peer Production Era." In van Abel (ed.), *Open Design Now*, 86–95. Amsterdam: BIS Publishers.

4

DIWO TO DAOWO

Rehashing Proprietorial Dominance of Art Practice

Ruth Catlow and Marc Garrett

Furtherfield, a platform for critical practices in arts and technology, coined the term DIWO (Do It With Others) in 2006, before the 'art hack' became a 'thing.' DIWO is a manifestation of grounded explorations and collaborations between networked peers, whose practices involve an open mixing of components from different sources, building new hybrid art experiences. This approach renegotiates power roles between artists and curators, audiences and participants to reform art worlds.

Since 2015 Furtherfield has brought DIWO approaches to critical engagement with the blockchain, smart contracts and cryptocurrencies. Ongoing exhibitions, labs and debates seek to explore how arts-led experimentation with organisational forms and governance might use Decentralised Autonomous Organizations (DAOs implemented as smart contracts across blockchain protocols) to realise Decentralised Autonomous Organisation With Others (DAOWO) as an experimental space for emancipatory, collaborative, networked art practices – to form a DAOWO.

This chapter presents some of the motives and philosophies behind Furtherfield's development of DIWO, starting with a critique of proprietorial dominance of art practice. It traces DIWO's connections with Do It Yourself (DIY) culture: art culture, networked and new media art, hacktivism and politics. It relates artistic and collaborative social hacks to historic movements such as Situationism, Mail Art, Fluxus and Cyberfeminism and shows how these influences might be remixed as DAOWO today.

On Proprietorial Dominance over Artistic Production

Proprietorial behaviours enact psychological and concrete forms of coercion and violence through the social infrastructures and belief systems that we inhabit.[1]

Proprietorial domination acts through the presumption of ownership over our psychic states of existence and the field of material objects we possess and use. This extends into and through our daily use of digital networks creating new states of dependency, reliance and addiction[2] (Beales 2017). The meanings of the words *proprietorial* and *proprietary* are closely linked. Proprietary refers to the possession, ownership or holding of exclusive rights to something, specifically an object (something owned by a specific company or individual for instance). In the computing world, proprietary is often used to describe software that is not open source or freely licensed. Examples include operating systems, software programs and file formats.[3]

The Cambridge Dictionary definition of 'proprietorial' is especially poignant, 'like an owner: He put a proprietorial arm around her.' It is this aspect of its meaning that we find particularly useful as it brings us directly to the concept of biopolitics elaborated by Michel Foucault, which asserts that governments regulate their populations through the organisation of human bodies, through biopower.[4] Hardt and Negri developed Foucault's ideas, saying 'Biopower is a form of power that regulates life from its interior, following it, interpreting it, absorbing it and rearticulating it'[5] (Hardt and Negri 2001). Global digital infrastructures now mediate a deep psychological bias that asserts the right of patriarchal power to own our personal and social contexts without taking responsibility for the harmful effects of its influence on the environment.[6]

The art world provides us with an excellent example of a closed-in, proprietorial system. To be clear, we are talking here about that art world that is made of markets, investments and speculation, where the dynamics are fed by a circuit of international media organisations, prizes, fairs, biennales, all underpinned by, and enmeshed with, the interests of established art institutions. The boundaries of this art world (or idea of this art world) are blurred, so almost anyone who calls themselves an artist is in some way defined and shaped by it. Responding to these conditions, artists can find themselves reinforcing the bars of their own cages, competing with each other to perpetuate their own and others' subjugation, by reproducing the 'system and its hierarchies when the actual distribution of aggregate benefits is skewed towards the top tier of the art world'[7] (Sholette 2017). This is not only the result of self-exploitation by individual artists eager for crumbs of acknowledgement and remuneration by unscrupulous art world power players. Gerald Raunig proposes that the art world manifests a set of conditions imposed by state apparatuses instigated through conservative values with a historiography that promotes processes of marginalisation. And so, artists with an emancipatory bent must constantly deal with the consequences of reductive

> conservatisms, such as rigid canons, fixation on objects and absolute field demarcations, activist practices are not even included in the narratives and archives of political history and art theory, as long as they are not purged of their radical aspects, appropriated and co-opted into the machines of the spectacle.[8]
>
> *(Raunig 2007)*

Anna Brzyski agrees, arguing that 'the language of the canon obscures the historic existence of multiple, temporally and geographically situated canonical formations'[9](Brzyski 2007).

The top tier of the art world continues to ignore the combined effects of their values and actions on the wider ecology, and in this they are the glamourous counterparts to others in the global professional classes, all locked into the maintenance of position and status, expressed through the things they own and aspire to control. This meshes with the wider world in which our narratives are dominated by a tiny group of media barons, working in league with (often corrupt or corrupted) heads of state to promote a Neoliberal hegemony. Together they typically promote an atomisation and isolation of individuals in society by delegating responsibility for decisions, about what would make a good life, to markets and technologies. At the same time, they game the system to maintain their power by crushing any alternative that challenges imposed top-down agendas.

So the question we have grappled with since our first collaborations with artists, techies and activists across the Web in the mid-90s is how to break the chain of submission within the proprietorial absolute?[10] How instead do we build interdependent, semi-autonomous, semi-permanent places, spaces, resources and contexts, to grow more various alliances, perspectives and interests? What tools and knowledge can we apply, individually and collectively, from exploratory, peer 2 peer, artistic, technological and cultural practices? How do we cultivate collaborative practices that take account of contemporary societal and environmental realities and reach out to others to do the same?

Furtherfield's collaboratively constructed platforms and spaces, in digital and analogue modes, represent the values of its user base, symbolically and practically. The spirit and ideas of Pink and Situationism have strongly influenced and shaped the organisation's identity. This is emphasised through Furtherfield's ongoing dedication to building independent art platforms and DIY culture with Free and Open Source Software. It's community of artists, techies, reviewers and discussants have initiated the building of online and physically collaborative, self-produced, cultural platforms since 1996; which has always involved negotiating tensions and ambitions, between its 'users and the extractors of value, visibility and invisibility.'[11] (Gere 2004). In particular, it has asserted the function of art as an emancipatory initiation or catalyst for social change and has sought to bypass the dominant idea of both the singular genius artist and art solely as commodity.

Don't Just Do It Yourselves Do It With Others!

Furtherfield coined the term *DIWO* in 2006 as an update of the original DIY spirit of punk and early Net Art, which eschewed notions of craft, instead using instruments and tools in the raw to take and make the stage on their own terms. A practical and cultural evolution, *DIWO* invites anyone with access to the Web, to take, play and collaborate with others using and building tools across networks, digital infrastructures and platforms. A critical dimension challenges traditional art making, and in particular, infrastructural and established art hierarchies. While

> Net Art in itself is not inherently communitarian even if the internet is a natural fit with 'radical cooperation' and distributed authorship. [...] Rather than perpetuating a modernist situation where artists' egos created a hierarchy, Furtherfield experimented with breaking down individual identities within a group.
>
> *(da Rimini 2010: p. 191)*

> Peers connect, communicate and collaborate, creating controversies, structures and a shared grassroots culture, through both digital online networks and physical environments. Strongly influenced by Mail Art projects of the 60s, 70s and 80s demonstrated by Fluxus artists' with a common disregard for the distinctions of 'high' and 'low' art and a disdain for what they saw as the elitist gate-keeping of the 'high' art world...[12]
>
> *(Furtherfield 2007)*

On 31 January 2007, Furtherfield made an open call to the first official DIWO exhibition *DIWO: E-Mail-Art,* first via its own community email list *Netbehaviour,* and then to other media art lists and various art groups. All 900+ subscribers to the Netbehaviour email list automatically became a part of the month-long event, receiving content directly to their email inboxes. This included everyday correspondence, instructions, code poetry, software experiments, remote choreography, remixing and tool sharing. Participants worked across the Internet, together across time zones, 'geographic and cultural distances with digital images, audio, text, code and software. They worked to create streams of art-data, art-surveillance, instructions and proposals in relay, producing multiple threads and mash-ups' (Catlow and Garrett 2012).

Over a hundred artists participated in the physical exhibition that took place at Furtherfield's *HTTP Gallery,* London, between 2 March and 1 April 2007. It was co-curated between all the contributing artists and Furtherfield in an event retroactively named *Curate With Others (CWO).* While they conformed to the exhibiting rules of being showable in the white cube gallery, the array of art forms still disrupted tradition by extending beyond their physical object status, existing as connected, decentralised actors influenced in real-time by a networked space beyond the gallery walls.

The DIWO graphic has been regularly updated since 2006 to visualise its cultural, social and technical contexts. It is important in that it acknowledges the many non-human actors involved in an ecological or cybernetic feedback process – with category-hopping networks of actors involved in artistic production: tools; technical devices and functions; semiotic elements and material infrastructures.

This version of the image depicts:

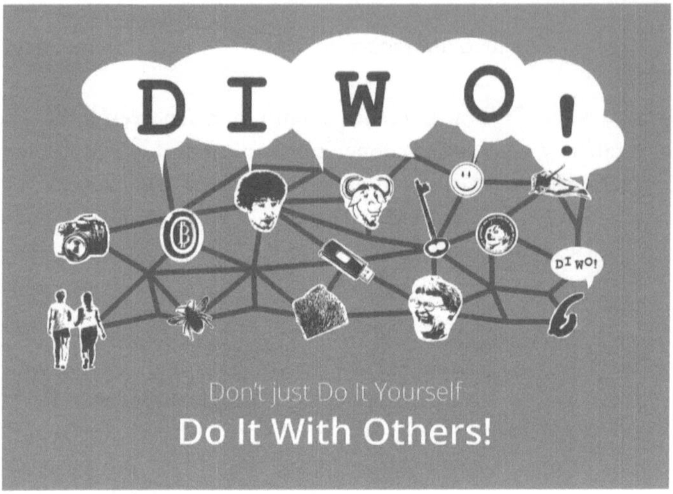

FIGURE 4.1 DIWO Graphic, Furtherfield (2015) (updated from original 2006 graphic).

TABLE 4.1 Description and relevance of icons in DIWO Graphic (Figure 4.1)

The Camera	A tool for recording the visual world.
The Bitcoin	Digital currency that allows people to attach a value to and exchange digital assets. Based on Blockchain.
The Philosopher	Contemplation of the nature of the world around them.
The Gnu/Linux logo	A community of values for Free and Open Source Software production.
The Emoticon	The language of signs and symbols, to concisely convert and communicate emotion.
The Drawing Hand	The element of making and hand crafting. It represents the bridge between the analogue and digital.
The Social Grouping	People, couples, groups, communities, societies. This encompasses personal familial and political relations.
The Insect	To represent all non-human living beings.
The Tuft of Grass	This symbolises the grassroots of organisation and production based in grounded experience. It is also a link back to earlier forms of grassroots activism such as *The True Levellers* and *The Diggers*.
The USB Stick	The storage and distribution of information.
The Key	Security and keeping private what needs to be private, PgP and encryption.
The DogeCoin	Decentralised forms of networked play. Memes, tribalism, and play with symbols and language.
The Hysteric	Represents the human nervous system plugged into the net. Contagion, affect, feeling.
The Talking Dildo	The libidinous energies flowing through and forming the net. Pornography and sexual exchange were a major driver and economy for Internet technologies.

After the first DIWO event, we wrote the DIWO Manifesto, as a guide for those interested in exploring its ideas and enacting its processes.

It's DIWO If It...

- Enlarges artistic freedoms.
- Uses the metaphors, tools, cultures and processes of digital and physical networks.
- Is led by experimental artistic processes rather than utilitarian or theoretical concerns.
- Disrupts traditional hierarchies and concepts of ownership working with decentralised peer 2 peer practices.
- Involves diverse participants (unwitting and active collaborators), ideas and social ecologies.
- Generates unruly and provocative relationships between symbolic meanings and material effects.
- Co-creates a new, freer, art context for more and more diverse people.

(Furtherfield n.d.)

DAO With Others!

DIWO grew out of the DIY, punk and post-punk, and Internet cultures of the nineties, and early noughties. *DAOWO* responds to what happened next to the Internet, to money and to art. By 2006, the colonisation of the Web by the profit logic of the 'big five,' served to sanitise communality and introduce new restrictions upon imaginative freedoms and collective intelligence across digital networks. By 2017, the contagion of rage-fuelled state-sponsored trolls who adopt tactical media and culture jamming techniques of earlier subcultures, amplify fascist tendencies in the most unexpected places, from the White House to the online silos of gamers, and identity activists (Nagle 2017). In 2008, coinciding with the financial crash, during which governments bailed out banks with taxpayers' money, Bitcoin, the first digital currency, was launched, underpinned by a new Internet protocol called the 'blockchain.' Meanwhile, art was being developed as a new financial asset class for investment, speculation and circulation by the super-rich. Across all zones of art-making, we now see the relationship between the value of arts to society and the way it is resourced and funded becoming increasingly awkward.[13]

We first started talking about DAOWO in Autumn 2015 with the launch of Furtherfield's *Art Data Money*, which aimed to draw an active international community of artists, technologists and activists to examine the possibilities for increased collaboration and sustainability in the arts offered by big data and the blockchain. We invited them to join us in a programme of exhibitions, labs and debates, online and at our two venues, a gallery and lab space in the heart of Finsbury Park in North London 'to build a commons for arts in the network age.'[14] We published a text by Rob Myers called *DIWO to DAOWO*[15] which inspired us to think of blockchains as a new context for DIWO. This text alerted readers to the extraordinary capacity of emerging blockchain technologies to carry the utopian dreams of widely divergent political positions (just as the early days of the Web had). Bitcoin was the first digital currency created and secured using cryptography across a decentralised network of users. Cryptocurrencies (there are now many of them) delegate trust and responsibility for global money flows to the users of cryptographically secured networks of computers, removing (so the rhetoric goes) potentially corruptible human institutions from global finance. With the introduction of a new programming layer to blockchains, in 2013 by Ethereum, this decentralisation extended to the delegation of global governance.[16]

> DIWO (Do It With Others) is a distributed campaign for emancipatory, networked art practices instigated by Furtherfield in 2006.
>
> A DAO is a Decentralised Autonomous Organization, effectively a corporation or a charitable trust implemented in networked computer code.
>
> Both are decentralised. A DAO has no single point of failure on the network, existing on the blockchain. A DIWO event is organised online and open to participants worldwide.
>
> Both are an application of network technology to social organization. A DAO implements a corporation, charity, club or cooperative. A DIWO event organizes artists, curators and writers to produce and exhibit work together.
>
> Both are rule driven and participatory. A DAO consists of trustless, incorruptible code that serves the interests of its members or clients. A DIWO event is a themed open call for artworks.
>
> Both are ways of managing scarce resources. A DAO manages resources such as a cryptocurrency token or the lock on a door. A DIWO event allocates the productive efforts and attention of an audience and the display space of a gallery.
>
> A DAOWO would be a combination of the two Decentralised Autonomous Organization With Others.[17]
>
> *(Myers 2015)*

We set out to question ourselves, to look afresh at relationships between aesthetics, governance and money through the protocol that promised to automate every online interaction as a transaction. We wanted to know whether blockchain technologies could platform and support the production of experimental prosocial art-tech processes, practices and experiences; thereby increasing access to new ways of thinking that would work for people, environment and society; bridging business and the arts to critically explore relationships between art and finance.

We faced a number of unexpected challenges. Blockchains, smart contracts and cryptocurrencies are surrounded with a hype hardly seen since the arrival of the Web. However, it turned out that in 2015, the tools, vocabularies and infrastructures of this technology were at a much earlier stage of development that was suggested by the rhetoric and the proliferation of blockchain start-up websites online. Blockchain vapourware was everywhere, but there existed very few actual functioning tools or platforms.[18] Unlike the Web in the mid-90s, when anyone with access to the Internet – artists, developers, any enthusiast – could code a little html and serve their pages to each other around the globe, blockchain requires a much deeper level of programming nouse. This puts the process of intuitive play and exchange with tools that had been such a central feature of DIWO events, out of reach for most people.

In the process of making our short film, *The Blockchain – Change Everything Forever* (2016),[19] we spoke to artists, developers, activists, business people and theorists working in the space and discovered that blockchains were at the same stage of development as the Web in the late-80s. We also discovered grave concerns about the potential for the code (of smart contracts) to effectively become law, defined by a tiny group of developers, funded by wealthy venture capitalists, operating outside of regulation, overturning centuries of evolution of global governance and law.

We can add to these issues the, as of yet, unresolved technical obstacles of scalability and environmental cost of the Bitcoin protocol; the association in most peoples' minds with the spectacular get-rich-quick culture of cryptocurrency speculators. Also irksome for many of the artists who grew up with the digital abundance of the WWW is the return by blockchains to digital scarcity and the reintroduction of techniques of ownership into digital network spaces. The business of art is now funnelling massive investment into the blockchain space in the areas of IP, provenance, fractional ownership and investment, advancing a deeper and wider financialisation of art worlds than seen before.[20]

However, we see opportunities gradually opening up for collaborative platform-building by-and-for communities of experimental artists (in the expanded sense of the word), participants and audiences who want to create not just saleable or tradeable art objects but to extend and diversify art contexts that include questions of organisation and governance. These and many other aspects of blockchain affordances and cultures have been explored through two exhibitions – The *Human Face of Cryptoeconomies (2015)* and *New World Order (2017)* –, our book *Artists Re:Thinking the Blockchain*[21], an art commission, articles, and the *DAOWO* workshop series devised and run with Ben Vickers (Serpentine Galleries)[22] which employs a range of experimental participatory processes from LARPing to Theatrical improv and hot-seating.

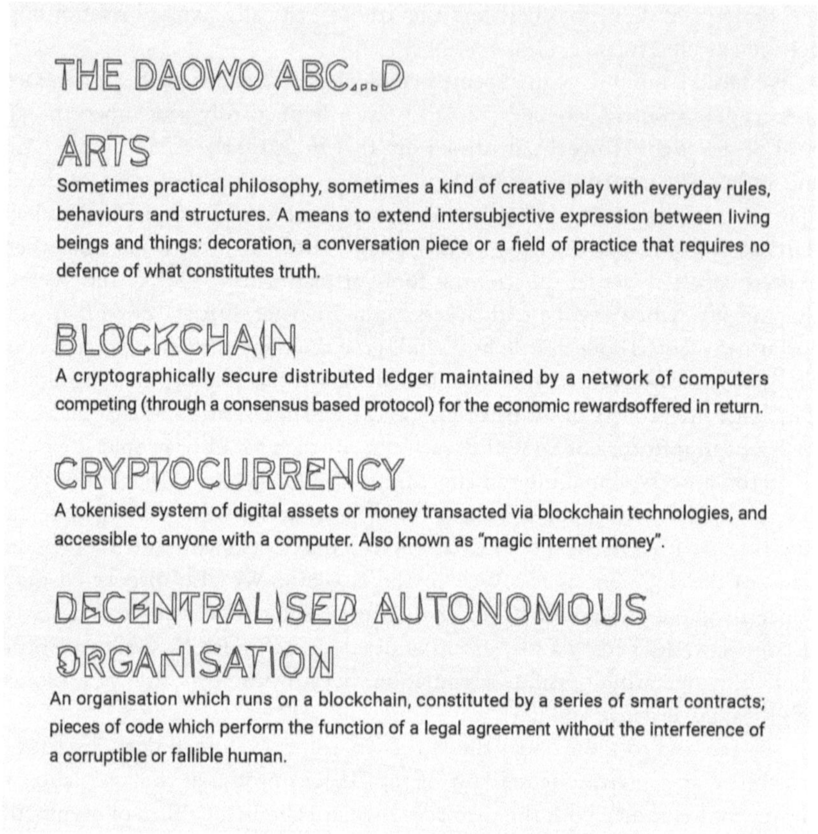

FIGURE 4.2 Illustration by Maz Hemming, created as part of the Live Action Role Play, *What will it be like when we buy an island (on the blockchain)?* For the DAOWO blockchain laboratory and debate series for reinventing the arts.

Artists are now widely at work in the creation of blockchain-native critical artworks like *Clickmine* by Sarah Friend[23] and *Breath (BRH)*[24] by Max Dovey, Julian Oliver's cryptocurrency climate-change artwork, *Harvest*[25] and *2CE6...* by Lars Holdhus,[26] to name but a few.

Artistic projects such as *Terra0*,[27] the self-owning, self-exploiting forest, and *Plantoid* by O'khaos,[28] the autonomous blockchain-based artwork that reproduces itself, 'harnessing the power of beauty and automated governance,'[29] both offer examples of blockchain-based governance systems that invite us to critically 'imagine a world in which responsibility for many aspects of life (reproduction, decision-making, organisation, nurture, stewardship) are mechanised and automated.'[30] Both artworks demonstrate functioning systems and help us to think through how we might determine and distribute artistic (and other) resources, their value and the rules for their co-governance, for the kinds of freedoms, commonalities and affiliations that are important for the arts.[31] Importantly and controversially, artworks like *Terra0* and *Plantoid* propose to deal with the problem of corruptible institutions by replacing their proprietorially inclined humans with algorithms. They both help us to think through the implications of a politics of automation.

> Powerful technologies develop to reflect the interests and values of those who develop them, but impact the everyday lives of us all.
>
> *(Catlow 2017)*

This text takes us through some of the artistic peer practices employed in the process of rehashing proprietorial dominance after the Internet. The power to create our own contexts is constantly under threat by those who would lock down history, territories, systems, places, spaces and consciousness, for their own narrow interests. If we ignore the effects of emerging technologies, we will become its victims. We are still in the early days of establishing a functioning platform and context for DAOWO, but realised full strength, it will underpin new grass-roots artistic ecologies and economies.

Notes

1 This section develops on a post made for the Transmediale *Face Values* PhD workshop *Unlocking Proprietorial Systems* by Marc Garrett (2018) https://researchvalues2018.wordpress.com/2017/12/18/unlocking-proprietorial-systems/ [Accessed January 28, 2018].

2 Beales, Katriona. (Oct 22, 2017) Are We All Addicts Now? An interview with Katriona Beales. Interview by Ruth Catlow. Furtherfield. Available at: https://www.furtherfield.org/addicts-now-interview-katriona-beales/ [Accessed July 28, 2018]

3 Garrett, Marc. 2017. "Unlocking Proprietorial Systems for Artistic Practice." *Research Values* (blog). December 18, 2017. https://researchvalues2018.wordpress.com/2017/12/18/unlocking-proprietorial-systems/ [Accessed January 28, 2018].

4 Ibid.

5 Hardt, Michael, and Antonio Negri. 2001. "Biopolitical Production." In *Empire*, 23–24. Cambridge, MA: Harvard University Press.

6 Garrett, "Unlocking Proprietorial…"

7 Sholette, Gregory. 2017. "Bare Art, Debt, Oversupply, Panic! (On the Contradictions of a Twenty-First-Century Art Education)." In *Delirium and Resistance: Activist Art and the Crisis of Capitalism,* Kim Charnley (ed), 65. Pluto Press. Available at: doi:10.2307/j.ctt1n7qkm9.11 [Accessed January 18, 2019].

8 Raunig, Gerald. 2007. *Art and Revolution: Transversal Activism in the Long Twentieth Century*. Translated by Aileen Derieg. Los Angeles : Cambridge, MA: Semiotext: 19.

9 Brzyski, Anna, ed. 2007. *Partisan Canons*. Edited by Brzyski, Anna. Jensen, Robert. Elkins, James, Cutting et al. Durham, NC: Duke University Press: 7.

10 Ibid.

11 Gere, Charlie, 2004. "New Media Art and the Gallery in the Digital Age – Tate Papers, no. 2." *Tate*. Available at: https://www.tate.org.uk/research/publications/tate-papers/02/new-media-art-and-the-gallery-in-the-digital-age [Accessed: May 3, 2017].
12 Catlow, Ruth, and Marc Garrett, 2012. "DIWO: Do It With Others – No Ecology without Social Ecology." First published in *Remediating the Social 2012*. Simon Biggs (ed), 69–74. University of Edinburgh.
13 This text draws on our ongoing programmes and debates described in *Furtherfield Spring 2018 Editorial: Blockchain Imaginaries*. Available at: https://www.furtherfield.org/blockchain-imaginaries/
14 http://www.furtherfield.org/artdatamoney/
15 Myers, Rob. 2015. "DAOWO." *Rob Myers*. September 13, 2015. https://robmyers.org/2015/09/12/daowo/.
16 It is ironic, given the spectrum of libertarian to anarchist spirit of early blockchain, just how much the space is now dominated by development in banks, governments and international big business. Catlow, "Are We All Addicts Now?"
17 Myers, "DAOWO."
18 In the Collision arts residency undertaken by Ruth Catlow and Kei Kreutler at London Southbank Centre.
19 Concept and Research by Ruth Catlow, Directed by Pete Gomes https://vimeo.com/184677819
20 Elhanani, Zohar. 2018. "How Blockchain Changed The Art World In 2018." *Forbes*. December 17, 2018. https://www.forbes.com/sites/zoharelhanani/2018/12/17/how-blockchain-changed-the-art-world-in-2018/#4ab771133074 [Accessed January 19, 2019].
21 Garret, Marc, Ruth Catlow, Sam Skinner, and Nathan Jones (eds), 2018. *Artists Re:Thinking the Blockchain*. Liverpool: Liverpool University Press.
22 In partnership with Goethe Institut-London.
23 Clickmine, 2017, Furtherfield and NEoN commission as part of the State Machines EU cooperation programme https://clickmine.click/
24 *Breath (BRH)* http://maxdovey.com/?page=performance&id=respiratory-mining
25 *Harvest* https://julianoliver.com/output/harvest
26 Myers, Rob, 2017. "You Can't Hodl This Holdhus." *Furtherfield*. December 11, 2017. https://www.furtherfield.org/2c-e6-85-dc-0b-a7-9f-2c/.
27 http://terra0.org/
28 http://okhaos.com/plantoids/
29 Ibid.
30 NEW WORLD ORDER Exhibition: https://www.furtherfield.org/wp-content/uploads/2017/04/NEW-WORLD-ORDER-Exhibition-Catalogue.pdf
31 The relationship between cryptocurrencies and governance is highly controversial.

> The Decred cryptocurrency has put governance front and centre. As well as moving to a hybrid Proof of Work / Proof of Stake system it has implemented an 'on-chain-governance' system. Decred contains the forum for its own critique and transformation, implemented as an extension of the staking and voting system used by its Proof of Stake system.
>
> Myers, Rob. 2018. "Blockchain Geometries." *Furtherfield*. January 22, 2018. https://www.furtherfield.org/blockchain-geometries/ [Accessed February 1, 2018]

Meanwhile FairCoin (another cryptocurrency being rolled out by FairCoop with the Catalan Integral Cooperative) ties its value to new forms of decentralised social organisation on the ground.

Bibliography

Beales, Katriona. (Oct 22, 2017) Are We All Addicts Now? An interview with Katriona Beales. Interview by Ruth Catlow. Furtherfield. Available from: https://www.furtherfield.org/addicts-now-interview-katriona-beales/ [Accessed: 28th July 2018]

Catlow, Ruth and Garrett, Marc (2012) DIWO: Do It With Others – No Ecology without Social Ecology- Essay. First published in Remediating the Social 2012. Editor: Simon Biggs University of Edinburgh. Pages 69–74.

Catlow, R. (2017) Introduction. *Artists Re:thinking the Blockchain* eds Catlow, R., Garrett, M., Jones, N. and Skinner, S. Torque, UK. London. Pages 21–37.

Da Rimini, Francesca. (2010) Chapter Four, Furtherfield. Socialised Technologies, Cultural Activism, and the Production of Agency. P.191 Available from: https://opus.lib.uts.edu.au/handle/10453/36069 [Accessed: 28th July 2018]

Furtherfield (2007) Do It With Others (DIWO) E-Mail-Art. Open call on the Furtherfield Website. Available from: http://archive.furtherfield.org/exhibitions/do-it-others-diwo-e-mail-art [Accessed: 28th July 2018]

Furtherfield (n.d.) Furthefield Webpage. Available from: http://archive.furtherfield.org/projects/diwo-do-it-others-resource [Accessed: 28th July 2018]

Nagle, Angela. (2017) *Kill All Normies: The Online Culture Wars From 4chan and Tumblr to Trump and the Alt-right.* Lanham: John Hunt Publishing.

+ CONTEXT II

Labs and Fab Labs

5

PERVASIVE MEDIA STUDIO

Propagating Practice

Clare Reddington

In 2008, Watershed launched the Pervasive Media Studio – an arts and technology research space designed to challenge the temporality, mono-culture and output-driven thinking of project funding and weekend hacks. By co-locating a network of artists, creative companies, technologists and academics, we set out to enable the exploration of emergent ideas, experiences and applications over time.

Today, the Studio has an international reputation for collaborations crossing cultural, commercial and academic sectors. Whilst technology platforms,

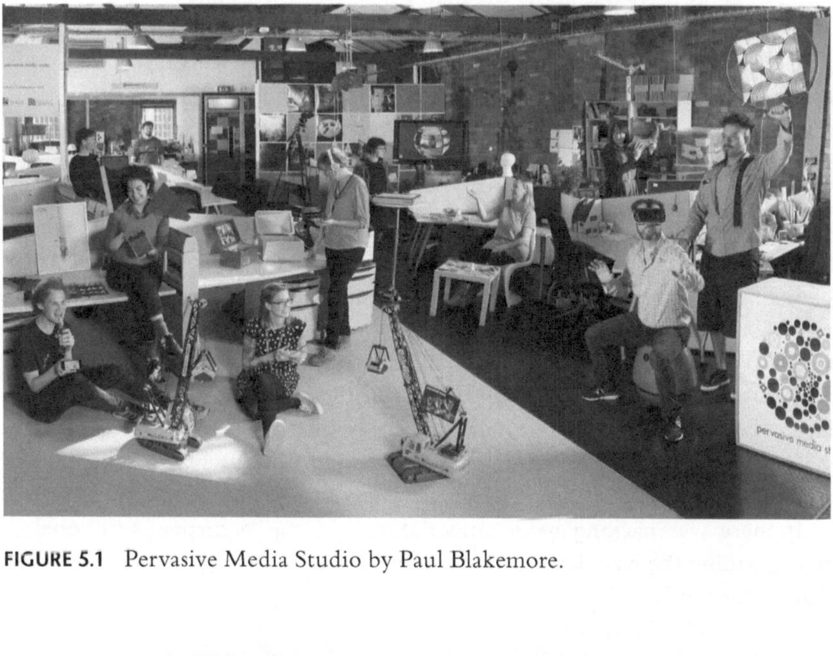

FIGURE 5.1 Pervasive Media Studio by Paul Blakemore.

reach, profile, community and methods have evolved and changed, an applied, values-driven, inclusive approach has remained at the heart of Studio culture.

This chapter shares some of the ideas that informed our thinking around the development of the Pervasive Media Studio and explores what we appropriated, experiences that changed us and what we have grown and changed.

Background

Established in 1982, Watershed[1] has played a key role in the development of moving image culture across Europe. Based in Bristol, UK, Watershed is recognised for its internationally distinctive programme of invention and talent development; as a leading centre for film culture and as Bristol's city centre cultural meeting place of choice.

In the 1990s, as the internet began to transform the world, Watershed began to explore its application through collaborative research with HP Labs:[2] on projects including SE3D[3] and Mobile Bristol.[4] Through a decade of shared research, Watershed, HP Labs and partners began to understand and articulate the power of inviting artists to use cutting-edge tools before they were readily available – providing real-world tests of their potential, as well as developing design principles and UX guides for their application.

The impact on artists was evident – they became early-adopters, communicating and questioning their use (and ethics) to a wider audience. The benefits of shaping and supporting a cohort of practice around a specific area of R&D was also clear, with participants sharing learning and mistakes as they happened – short-cutting the collective innovation journey.

The Pervasive Media Studio was founded by Watershed and HP Labs[5] to leverage this learning and was opened on 14 February 2007 by Alistair Darling, the new Chancellor of the Exchequer. That we didn't want a launch nor have anything concrete to announce was immaterial; "What the Chancellor opens does not close" said the Regional Development Agency (who did not wear so well).

When we picked up the keys to Leadworks (the Studio's first home), we perhaps had a clearer idea of what we were not rather than what we would become. There were (and are) many brilliant media arts labs around the world (lots of them are documented in BALTAN's Future of the Lab[6]), but we weren't creating one of those – we wanted something purposefully curated and co-located within both cultural and commercial propositions. We wanted to be open to creative practitioners from all art forms, and to both institutions and independent practitioners.

A place that felt open to everyone (including those who couldn't code), that valued process as well as output, that moved beyond time-bound projects – to make a space where collective wisdom and skills could be built on, circulated and renewed, rather than ebb away when time or money ran out (as had happened repeatedly).

If there was hacking being undertaken – it was perhaps institutional – re-examining the way that artists are supported and funding is applied – rather than of the work itself.

Today, the Studio is a collaboration with UWE Bristol and University of Bristol and is comprised of a thriving community of creatives, academics, technologists, start-ups and industry, exploring early-stage ideas in the sphere of creative technology. We offer fixed desk space, hot desking and meeting and event space. We have been home to hundreds of residents and thousands of others have passed through, to share learning, ideas and questions to build a thriving network of practice.

Things We Appropriated

Open innovation describes a mindset that seeks to burst out of traditional silos and institutional thinking to connect with ideas and knowledge from external parties. It champions an ecology of ideas, where wisdom is openly circulated, shared and built upon.

A major driver of the Pervasive Media Studio is articulated in John Hagel and John Seely Brown's Creation Nets,[7] where difference is understood as an asset. In times of rapid change, making new ideas requires collaboration with actors such as skills, culture, size or outlook:

> Creation nets accelerate innovation across participants. Not only are participants able to innovate more rapidly than they could outside these networks, the pace of their innovation accelerates the longer they participate in the network[8]

With existing access to the universities and to researchers at HP Labs, Orange and other corporates, our early focus was on identifying individuals and start-ups who wanted to make locative work. A few residents were tempted in with access to facilities like hot showers and a kitchen; most were working out of their garages, studios and bedrooms and looking to harness the power of joining something bigger than themselves. Producing a community gave us a bigger story to tell – and through a mixture of stunts,[9] events and research,[10] profile was gained for people whose work often didn't yet have a market, let alone a route to it.

Hagel and Seely Brown describe the crucial input of a network organiser in any Creation Net – people who enact a 'distinctive set of management techniques to ensure focus and value creation.' For us, this meant investing in Producers, with the ability to curate a community above the curation of artworks or products. Curating the Studio, then and now, involves ensuring a balance of cultural and commercial, technology and experience-design individuals and companies. When a person applies to join the Studio, their idea could be excellent, but if there are already a lot of residents from a similar discipline, the timing might not be right for them or for the community.

> Being around people from such different backgrounds made me realise that you don't have to give up if you can't do something, you can dream

> as big as you like because there will always be someone who knows how to do something that you can't.
>
> *Tenaya Steed, Pervasive Media Studio New Talent Resident 2016*

Another version of openness we appropriated was from Jelly, a casual co-working event[11] that first took place in New York homes, workspaces or coffee shops. In Jelly's own words:[12]

> We provide chairs and sofas, wireless internet, and interesting people to talk to, collaborate with, and bounce ideas off of. You bring a laptop (or whatever you need to get your work done) and a friendly disposition.

Replicating Jelly's informality and openness, Open Studio Friday[13] has been running since our first year. Fridays are days when anyone can work in Pervasive Media Studio, connecting them to new ideas, offering them a taste of Studio life and introducing us to new talent. Every week our community grows by 15–50 people, having tours, getting down to work, and attending our lunchtime talk. In turn, by sharing our learning as openly as we can, we have a sister Studio (Kalieder in Exeter[14]), and our methodologies have informed other cultural organisations across the world.

Things that Changed Us

Sustaining openness relies on embedding a sense of fairness, shared value and equity within the structure of community interaction and support. Early on we decided to gift desk space to Studio residents instead of charging them – to ensure we could maintain difference in who we are able to support. The Creation Net would soon unravel if affordability became a barrier to entry.

In Bill Sharpe and Graham Leicester's *Producing the Future*,[15] a study of Watershed's Role in Ecosystems of Cultural Innovation, they explore the benefits of putting the 'money at the margins' of the community 'We certainly need money to flow through the system, but not in a way that degrades the other currencies of value.'

Our understanding of the multiple currencies at play in the Studio was developed by Goetz Bachmann et al.,[16] who identified it as a triangular gift economy, where generosity is generalised across all players:

1. Producers give access to resources, in the form of workspace, support and overall access to the studio (including its prestige).
2. Residents return this gift partly by producing successful projects and partly by engaging in a collaborative working environment, which then is also a gift to other residents, adding to the value of the studio.
3. Both projects and collaborative environment produce material for stories, which the producers can use to acquire new funds.

The insight gained through working with UWE researchers on value circulation was profound, and we wanted to ensure this exchange was also made explicit to the community. One tangible example of this was including generosity and openness within the occupancy agreements that all Studio residents are asked to sign:

1. Interruptibility: at any time, we encourage members of the Studio community, Watershed staff and visitors to the Studio to interrupt one another and find out what the other is doing, ask questions, seek advice, share opportunities etc. By being open, curious and interruptible, we are all able to make better connections, share our expertise, and create the conditions for all of our ideas to be improved.
2. Generosity: the Studio thrives on the generosity of its community. We interpret this broadly, and it can encompass generosity of ideas, resources, running events, skill swaps or meet ups, sharing client leads and funding opportunities, collectively testing prototypes or just offering to make a cup of tea.[17]

These agreements ensure that shared value is enshrined from the outset, shaping the context, spaces and projects that people work within and deliver.

Of course, the need to leave money at the margins is in competition with the need to leverage funding for the community, a dilemma which showed itself during REACT.[18] In 2011, we were a partner in one of four AHRC Knowledge Exchange Hubs, which received £4 million to support collaborations

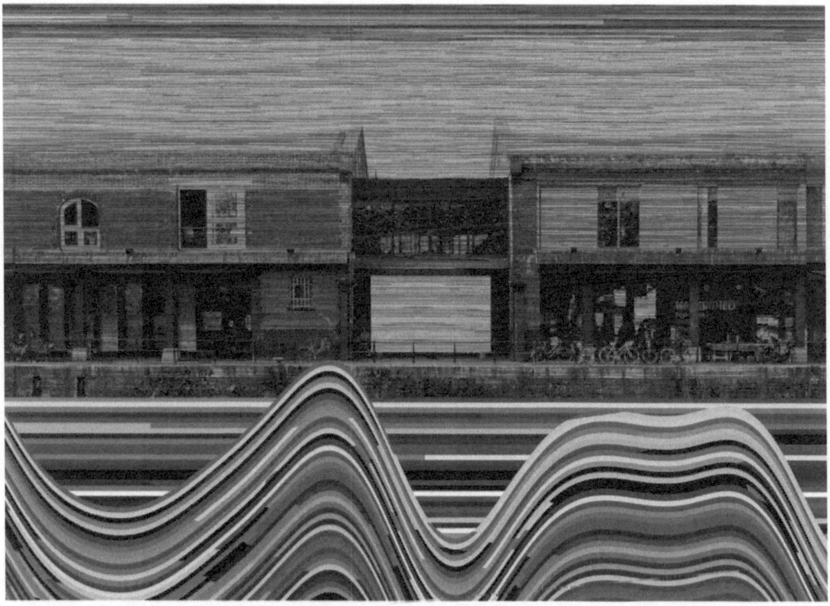

FIGURE 5.2 *Waveshed* by Joe Magee.

between academics and creatives. REACT scaled the reach and network of the Studio, contributed significant research around methodology and delivered many new people to our community. But, by the end of four years, we began to observe tensions around whether some residents were favoured with funding and profile more than others. In addition, the mix of Studio residents was weighted more heavily to the academy. This served as a timely reminder that moving money from the margins requires vigilance and care.

Things that Have Grown and Changed

A healthy ecosystem requires stewardship, and at the centre of Watershed's support for the Studio community are its team of Producers and its convening power to test and share in a real-world context.

Producers curate, support, question and champion the community – fostering 'a chaos of ideas, invention, imagination and possibility' and working with people to configure their ideas in 'ways that begin to gain cultural traction.'[19]

Producers are permission givers – instilling confidence in people with great ideas that they have the skills and ability to pursue them. They are translators, negotiating across sectors and within collaborations to find common language and ensure equitable relationships. They are opportunists; they spot potential and make introductions. Practically, Producers support Studio residents with advice, feedback and connections as well as identify funding and design and deliver projects. The capabilities and capacities of an effective Producer are covered in Kate Tyndall's *The Producers: Alchemists of the Impossible*,[20] and in Maureen O'Hara and Graham Leicester's *Dancing at the Edge*.[21]

For Watershed, the complementary backgrounds and skills of the producing team as a whole are also key to the Studio's continued renewal. Producers come from a range of backgrounds including visual arts, theatre, technology and TV production. This range of skills and networks enables us to design programmes with the widest possible benefit and to ensure that we remain highly connected. It has also allowed our recent focus on sharing this producing practice with others around the world through Creative Producers International.[22]

Working with talented people on early-stage ideas requires responsibility and commitment – to work with them over time, to develop language and taste and a shared set of possibilities. This is not just to the benefit of the individual – whilst the Studio certainly champions new talent – the Creation Net requires wisdom and experience to also be present to ensure maximum learning.

The Studio's values and culture are of course not fixed, and our energy and emphasis have shifted as the world has changed and the community matured. A recent focus is around growing our commitment to inclusion – as social responsibility as well as an innovation condition. The world of technology start-ups is still focused around the white boys of Hoxton.[23] It is vital that women, people of colour, disabled people, the LGBTQ+ community and people from low socio-economic background are not excluded. It is the only way to ensure the Creation Net remains productive.

The Studio has always attracted a significant number of female inventors (53% in 2017), a factor we understand as being about reflection (our team has also always been made up of a majority of female-identifying producers), as well as our language and welcome. Whilst we have made progress with allies like Unlimited,[24] and employed a Producer to make a specific invite to people from a BAME background,[25] there is still much to do.

One of the ongoing dilemmas of Studio care is in the balance of offering sustained support with ensuring the network does not become closed or cliquey. Techniques for countering this are centred around attracting new people rather than limiting the timeframe of our support. Telling our stories in many places, attending many kinds of networking events and encouraging Studio residents to also act as advocates and talent spotters, complement the methods described above to ensure a steady stream of new faces. An annual review process with each resident is structured to explore if they are still deriving value from their residency in the Studio and if they are similarly inputting to the whole. When the answer is no in either area, it is time to move on.

From year two, we instigated annual desk changes to ensure people did not get too comfortable in 'their space' and to mix up who sat next to who (and therefore the conversation). Whilst painful in the planning and execution, every move kick-started the energy within the community and resulted in renewed interaction.

In 2011, the Pervasive Media Studio left Leadworks and moved to a larger space in Watershed. This move resulted in more than renewed energy – the less linear desk set-up of the new space enabled a more fluid approach to working and proximity to Watershed bar increased social interaction across the residents. In addition, closer engagement with Watershed's Cinema and Engagement programme team resulted in collaborations such as Mail, Maps and Motion[26] (which combined live music and moving image on a huge scale) and the increased use of the Studio by the young people of Rife.[27]

Whilst early-stage testing of ideas had always been a key part of the Studio's approach, the rhythms and flows of entering, leaving and working in a public building (rather than a closed lab type space) changed both our working practice and our mindset.

With the use of the Watershed bar and conference spaces came increased opportunities for testing and showcasing. Closer working quarters with Watershed's communications teams created an engaged dialogue with a much larger audience. These shifts in resources and mindset resulted in more co-produced models of innovation like Play Sandbox, not only creating better products but also creating new models of engagement.[28]

> By playing the connector role, Watershed has developed to be a public-facing cultural and creative economy hub for the Bristol city region with strong networks amplified through the public programme. It is the open cultural approach combined with wide public engagement and a strong showcasing ethos which has enabled Watershed to leverage its profile to become a trusted connector.
>
> *Jon Dovey*[29]

The Studio has also benefited from Watershed's sustained commitment to inclusion, the serendipity of the many kinds of people who come through the building and its connection with the city, as both place of debate and civic leadership.

Whilst the decision to launch the Studio outside of Watershed was pragmatic (we had no space), it was also key to our ability to create new ways of working. Like Lookheed Martin's Skunk Works[30] (an internal innovation unit with its own rules and practices) we were able to break the rules and challenge systems around HR, meetings, communication, staff structure etc. However, this is only useful if we can also codify and meaningfully share learning. The Studio's move brought a renewed ability to embed new practice back into Watershed.

The Future

The Pervasive Media Studio was ten on 14th February 2018, and the community is thriving: a recent resident survey, undertaken with researchers from UWE Bristol, reported that the annual turnover of the Studio network is £13.7m, current residents are responsible for over 250 jobs and over 70% of respondents are delivering work internationally.

The Pervasive Media Studio is not a place of unquestioning technological optimism, nor one that encourages dystopian futuring. It is perhaps a place of scepticism – as defined by Studio resident Tim Kindberg in his recent talk series on technologies 'for Sceptics.'[31] To be sceptical requires an understanding and analysis of what is widely asserted to be true. Scepticism is both a questioning of agendas AND a creative response to them.

As the power structures and privilege of the culture, technology and media industries come under scrutiny – inclusive, accessible, welcoming spaces will ensure tomorrow's talent fulfill their true creative potential. Rather than waiting for the breadcrumbs of the technology industries to fall into the lap of the culture sector – we believe we should be a part of authoring the tools, the ways they are used and the narratives around them.

For as long as it is needed, the Studio will nurture open and generous practice, as it is the best way to change the world:

> Hope locates itself in the premises that we don't know what will happen. In the spaciousness of uncertainty is room to act.
>
> *Rebecca Solnit*[32]

Notes

1 Watershed website: https://www.watershed.co.uk/
2 Reddington, Clare, David Drake, and Erik Geelhoed. 2007. "Unusual Common Ground: The Watershed/HP Labs Partnership." *HP Laboratories Bristol*, September 7, 2007.
3 SE3D http://www.hpl.hp.com/se3d/se3d-intro.html
4 Pervasive mobile media project on magazine's "short list" for innovation prize, August 2004, http://www.hpl.hp.com/news/2004/july_sept/mobile_bristol.html
5 The team that founded the Pervasive Media Studio comprised of Clare Reddington and Dick Penny of Watershed, Phil Stenton of HP Labs and Bill Sharpe of Appliance Studio. We were quickly joined by Jon Dovey of UWE Bristol. The Studio was set up with cash support from HP Labs and the South West Regional Development Agency.
6 Baltan Laboratories, *The Future of the Lab*, August 25, 2010, https://issuu.com/baltanlaboratories/docs/tfotl_issuu
7 Hagel, John, and John Seely Brown. 2011. "Creation Nets: Harnessing the Potential of Open Innovation." *Journal of Service Science (JSS)* 1 (2): 27. doi:10.19030/jss.v1i2.4293 [Accessed April 6, 2018].
8 Ibid.
9 Waters, Darren, "Hunting the bowler hat at SXSW." *BBC New*, 14 March 2009, http://www.bbc.co.uk/blogs/technology/2009/03/hunting_the_bowler_hat_at_sxsw.html
10 The Pervasive Media Cookbook, DCRC Press, 2014, http://pervasivemediacookbook.com/
11 Grossman, Ann Jane. n.d. "Freelancers Forgo Office Space for Casual Coworking." *WIRED*, July 20, 2007. https://www.wired.com/2007/07/freelancers-forgo-office-space-for-casual-coworking/ [Accessed March 18, 2018].
12 Website: Jelly, http://www.workatjelly.com/
13 Website: Open Studio Friday, https://www.watershed.co.uk/studio/events/2016-04-29-1000
14 Website: Kaleider, https://kaleider.com/projects/studio/
15 Sharpe, Bill, and Graham Leicester. 2010. *Producing the Future*. International Futures Forum. https://www.watershed.co.uk/sites/default/files/publications/2011-03-15/Watershed_IFF_Report_online.pdf
16 Bachmann, G., J. Dovey, J. Monaco, and B. Sharpe. 2012. "Cultural Value Networks Summary Report." *Report or Working Paper*. September 26, 2012. http://eprints.uwe.ac.uk/18790/ [Accessed April 11, 2018].
17 Watershed Arts Trust's Pervasive Media Studio Memorandum of Understanding for Occupancy of Pervasive Media Studio as of January 2018.
18 REACT website http://www.react-hub.org.uk/
19 Sharpe and Leicester, *Producing the Future*
20 Tyndall, Kate. *The Producers: Alchemists of the Impossible, Arts Council England and the Jerwood Charitable Foundation*. September 30, 2014, https://issuu.com/jerwoodcf/docs/a11986_pl_the_producers_book_

21 O'Hara, Maureen, and Graham Leicester. 2012. "Dancing at the Edge: Competence, Culture and Organization in the 21st Century." October 2012. http://www.internationalfuturesforum.com/p/dancing-at-the-edge [Accessed January 19, 2018].
22 Creative Producers International Website: https://www.playablecity.com/programmes/creative-producers-international/
23 Torrance, Jack. 2015. "Silicon Roundabout's Boards Are Less Diverse than the FTSE 100." August 24, 2015. https://www.managementtoday.co.uk/article/1361232 [Accessed March 21, 2018].
24 Unlimited is an arts commissioning programme that enables new work by disabled artists to reach UK and international audiences. http://weareunlimited.org.uk/
25 Ash-Harper, Zahra. 2017. "What Is the Role of a Cultural Organisation during Black History Month?" *Watershed*. October 16, 2017. https://www.watershed.co.uk/articles/what-is-the-role-of-cultural-organisations-during-black-history-month [Accessed March 25, 2018].
26 Video: Mail, Maps and Motion, August 2012, https://www.watershed.co.uk/audio-video/mail-maps-and-motion
27 Website: Rife https://www.rifemagazine.co.uk/
28 Oldfield, Alison, and Helen Manchester. 2016. "Designing Playful Products: Nine principles for including children in collaborative, rapid R & D." *REACT.* http://www.react-hub.org.uk/sites/default/files/publications/0307-playsandbox_e_Paper.pdf [Accessed March 25, 2018].
29 Dovey, Jon, Simon Moreton, and Ian Hargreaves. 2016. "REACT Report, Celebrating the Value of Creative Networks." *REACT*, UWE Bristol. http://www.react-hub.org.uk/sites/default/files/publications/REACT%20Report%20low%20res_2.pdf [Accessed March 21, 2018].
30 Skunk Works website, Lockheed Martin, https://www.lockheedmartin.com/us/aeronautics/skunkworks.html
31 Event page: Blockchain for Sceptics, January 12, 2018. https://www.watershed.co.uk/studio/events/2018/01/12/blockchain-sceptics
32 Solnit, Rebecca. 2016. "'Hope Is an Embrace of the Unknown': Rebecca Solnit on Living in Dark Times." *The Guardian*, July 15, 2016, sec. Books. https://www.theguardian.com/books/2016/jul/15/rebecca-solnit-hope-in-the-dark-new-essay-embrace-unknown [Accessed January 12, 2018].

6

ELABORATING ON LABS

Reflections on the Blurring Boundaries between Arts, Science, Technology and Society

Olga Mink

Context

Baltan Laboratories[1] initiates free experimentation on the intersection of art, design, technology and science. By bridging the gaps between disciplines, we evoke inquisitive ideas and insights for emerging societal issues. The lab functions as a collaborative mindset, network and hub, connecting curious individuals and organisations. By placing art and design research at the core of its activities, Baltan explores the implications, promises and pitfalls of our technological society and translates these ideas into different outputs, including collaborations, events, workshops, publications and expos, creating a space for reflection, (practice-based) research and experimentation.

By curiously exploring the notion of what it means to be human, we look at how technology is changing our values, behaviours, perception, bodies, even our notion of reality. We believe that disciplines are containers created in an age of overspecialisation. By getting rid of these boxes, we bring together (un)related fields to create a larger body of knowledge. We don't have one methodology, we don't aim for pre-established outcomes, we don't think new is better. We speculate the imaginary, we focus on fundamental questions beyond the excitement of technology, we explore the present unknown by reclaiming the future and rethinking the past.

I write this at a time when the influence of digital technology on social, economic and artistic domains is dissolving existing borders between disciplines. It is now widely understood that new knowledge can no longer be derived solely from specialised, discipline-specific methods. However, while the fusion of knowledge can create new opportunities, it also confronts us with unforeseen challenges. Any attempt to solve these emerging challenges demands novel approaches, ways of thinking and the development and testing of new methods of working.

The technological transformations of the 21st century have also brought forth a compelling moment of cultural transition, which affords us the opportunity to redefine current values and rethink the meaning and role of art in tomorrow's society.

The Arts and Humanities constitute a domain that focuses on what it means to be human. Artists and designers engage in complex issues to which our current society has not yet found answers. Contexts such as labs, hubs and maker spaces do not shy away from these challenges emerging in today's society. We rise to these challenges by facilitating resources for interdisciplinary research and experimentation, developing hybrid forms of collaboration, providing access to (open source) technologies and facilitating unexpected connections. Failure is embraced as a valued part of our evolution into unforeseen or unknown directions. By establishing connections between organisations, industry, education, citizens and policymakers with artists and designers, we can create a breeding ground for artistic expressions which can form the basis of interdisciplinary and cross-sector models of co-production, cooperation and collaboration. These novel, experimental methods can aid the development of unconventional and human-driven approaches to existing challenges.

This chapter will focus on two hybrid research methodologies born within the context of Baltan Laboratories and shaped by its values. *Hack the Body*[2] and *Age of Wonderland*[3] are two initiatives that aimed to develop sustainable, practice-led innovation strategies connecting individuals, cultural organisations, high-tech institutes, international corporations, SMEs and NGOs. Hack

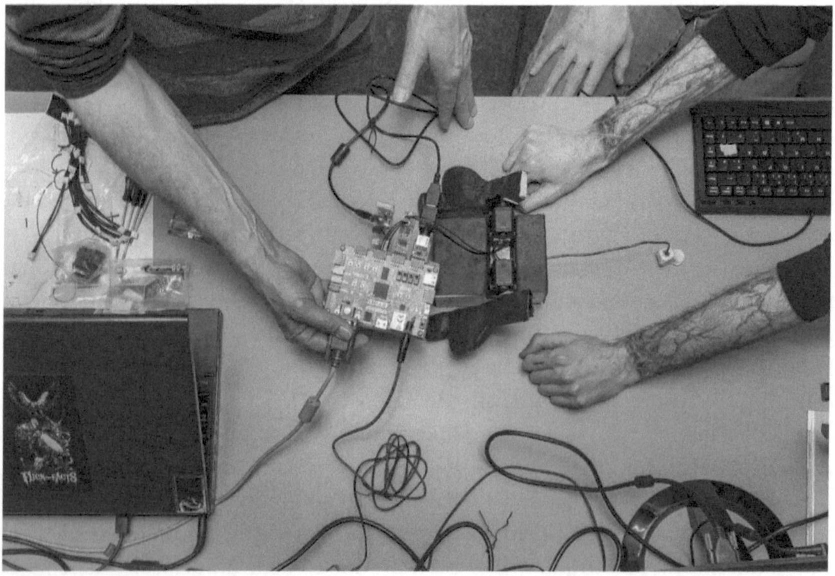

FIGURE 6.1 *Hack the body workshop* by Marco Dunnorumma. Photo by Sas Schilten.

the Body explores open innovation methods through artistic collaboration, while *Age of Wonderland* revolves around the investigation of social innovation for real-world challenges. New insights and learnings gained through the delivery of these two projects will be framed through an analysis of the impact of outcomes on project partners, users, audiences and Baltan itself, the models for experimentation and collaboration developed throughout each project's research trajectory and the context(s) in which the projects took place.

Hack the Body: Open Innovation through Artistic Collaboration

Open innovation involves opening innovation processes to society as a whole both by using knowledge and resources outside of one's own organisation and by making one's knowledge and resources available to society. Current methods for fostering open innovation include product platforming (APIs), idea competitions, challenges, hackathons, collaborative product design and development (user innovation). Baltan re-appropriates proven collaboration models such as open innovation and uses them to actively reach out to partners across different industries as they are invited into the open-minded context of our hybrid art lab. Within this context, we can act as a harbour for other organizations to explore their own themes and programs that fit the philosophy of the lab. We can connect cultural practices and values with industry, government and educational institutions and support them in their exploration of how they can leverage one another's values. Our goal is to foster long-term collaborations and support the generation of new insights.

Hack the Body is a multi-year open-innovation program that aims to connect partners from the creative sector and industry, in order to stimulate them to work together and develop projects around biometric data, the human body and identity. Working with partners, we reflect and explore the boundaries between the human body and wearable technology. We like to take a critical stance on our developing relationship and dependence upon data and sensors, and the amount of intimate data we generate from our bodies and our lives. These cultural approaches offer an alternative perspective compared to the traditional market-oriented technological R&D approach.

Artists, academics, scientists, engineers, companies and other cultural organizations are invited to participate in a program that offers a pool of expertise, knowledge, ideas, hardware and software that the partners involved can read, use, improve and further develop. *Hack the Body* brings together artistic projects that share the same underlying idea: using new sensor and information technologies to explore innovative concepts within biometric measurement, neuro-feedback and data generation. The program enables a diverse mix of disciplines to collaborate and participate in the generation of a range of outputs including artistic research, tools, events, open labs, publications, installations, performances, hackathons and artists residencies. This cross-disciplinary way of sharing new developments is a distinguishing factor of Baltan Laboratories. We believe that an open attitude towards co-creation and mutual inspiration is elemental in coming to new insights.

E.E.G kiss[4] by artists Karen Lancel and Hermen Maat is an example of a project that successfully emerged from the *Hack the Body* open innovation model. The project explores the idea of shared intimacy, data visualizations and sonication combining EEG technology. Lancel and Maat's work focusses upon whether a kiss can be measured and translated into data. The project responded to questions such as can we measure what kissers feel? Can we transfer a kiss and its intimacy online? By hosting live kissing experiments, visitors were invited to participate as kissers, voyeurs and E.E.G. data scanners. The installation consists of a love seat, with two opposite chairs, where kissers could engage. While kissing, their brainwaves were measured with E.E.G. headsets, made visible as E.E.G. data. Screens showed the data and a floor projection encircled the kissers with the real-time streaming data, as an immersive data cloud landscape. The soundscape was generated by the Brain Computer Interface, which translated the real-time E.E.G. data of kissing brains into a music score and algorithm for an E.E.G. kiss symphony.[5] Currently we are exploring the possibility of developing this project as a High-Tech Wedding ceremony with the same underlying technology. Lovers are united through E.E.G. technology by scanning their brainwave activity during the wedding kiss and by translating this data real-time into 3D printed wedding rings.

Age of Wonderland: Social Innovation for Real-World Challenges

Social innovation focuses on the process of innovation for emerging societal challenges[6] in which innovative ideas and human values are exchanged amongst stakeholders. To tackle difficult problems in technological society, the process of social innovation proposes an inclusive approach in which the non-profit, public and private sectors need to work together. Social innovation does not propose a strict methodology; rather it aims to incorporate a practice-based knowledge system that is openly shared amongst individuals and organizations. Even though wicked problems[7] are at the core of this methodology, social innovation acknowledges that a one-size-fits-all approach is not appropriate. Successful ideas are often strongly linked to the actors involved, the people who are in the lead and whether a community is open to change and to taking ownership of the problem. Creating a level of openness and trust between the group and the facilitators involved is an important key to the successful development of ideas into a community or specific targeted group.

Baltan devised *Age of Wonderland* program on the premise that we need a more robust collaboration between creative makers and thinkers from all over the world. We recognize that when it comes to securing a liveable, inclusive society and planet we need to actively embed the ability to innovate with and for local communities all over the world. The program invited creative innovators to develop ideas to boost sustainable social change on a global scale. Four editions of *Age of Wonderland* were developed, each addressing a different topic. The first edition in 2014 embraced the concept of creative practitioners as experience experts in *The friendly invasion of a new world order,*[8] proposing the idea that those who grew up with less resources may be much more creative and resilient. In 2015, the topic *Fair and Green Food*[9] dealt with the wicked challenges within our globalized food system, such as waste management, renewable energy or learning from nomadic knowledge and traditions in Kyrgyzstan.[10] In 2016, the topic *Big Data? Big Dada!* invited artist and designers to critically approach the use of big data in society, dealing with community building, ownership of data or data for predictive policy. The final edition in 2017 embraced the idea of (personal) knowledge dissemination with *100 Days of Learning.*[11] All projects developed were presented to an international audience during the annual Dutch Design Week in Eindhoven.

An example of a project that aimed to transform our perspective is the *Death Tolls Experience*[12] by Ali Eslami (Iran). It was presented during the *Big Data? Big Data!* edition. *Death Tolls Experience* explores the psychological effect of the

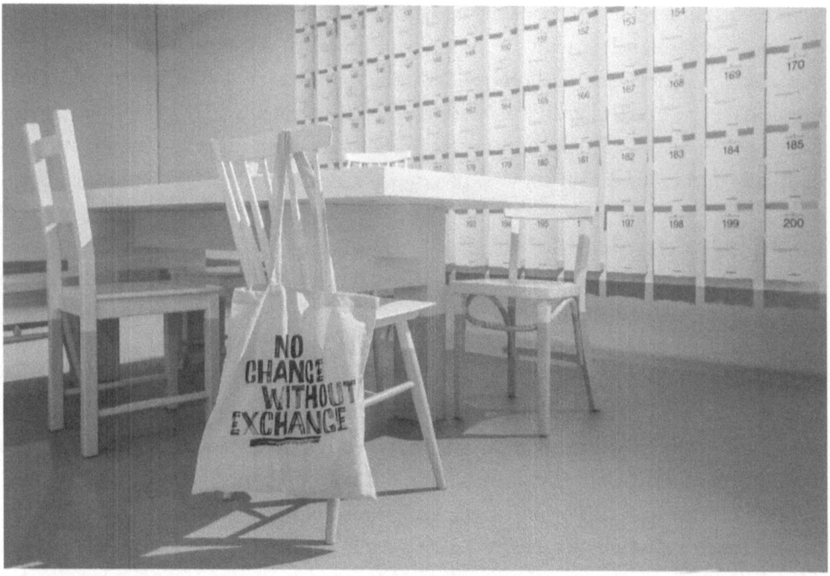

FIGURE 6.2 *Age of Wonderland, 100 Days of Learning.* Photo by Chris de Zeeuw.

mass media on our ability to understand and accept mass deaths. Whether they occur during the civil war in Syria or during the terrorist attacks happening in Paris, Brussels and in cities around the world, when we are bombarded with facts and figures all day, every day, they gradually lose their emotional urgency. Eslami uses virtual reality to investigate how this technology can not only restore empathy but also provide a context to news reports in which the numbers of victims can be counted but not comprehended.

Age of Wonderland enabled us to facilitate tailor-made collaborations between creative innovators of the global south and north. Over the course of four years, the program hosted international exchanges between creatives from Africa, Asia and Latin America with our local communities based in the city of Eindhoven. Collaborations emerged via a facilitated series of residencies, presentations, local follow-ups, knowledge production, research and development that helped to combine knowledge, resources and local networks, to create a lasting impact on distributed communities. We also sought to facilitate a legacy for these projects by creating new connections with start-up investors to support the implementation of outputs and potential solutions within local communities in Africa, Asia and Latin America upon the participants' return home. We found the exchange also encouraged social innovation in the Netherlands, as new insights towards problem-solving strategies were developed through collaboration with international creatives.

Impact

Through the delivery of *Hack the Body* and *Age of Wonderland*, Baltan aspired to enhance artistic and process-driven research and development by testing open and social innovation methods within the context of a hybrid art lab. Both methods leveraged individual as well as collective learning, by bringing together different networks of people and organisations to share ideas and foster interdisciplinary and cross-sector models of co-production, cooperation and collaboration. Each trajectory was deeply interwoven with the topics addressed in each of the programs. For example, within the *Hack the Body* open innovation model, we dedicated curated meetings to a limited number of topics, such as the use of sensors and algorithms in related projects or we explored the concept of synchronicity through emerging wearable technologies. These concepts would allow for a specific (artistic, scientific or academic) angle and inspire participants, as well as a broader field of practitioners from different fields. Consequently, these mediated events could accumulate in more targeted collaborations in which Baltan would play a significant role as mediator, facilitator, co-producer or co-financer.

The impact and potential of *Age of Wonderland* was most significant in the final edition in which we invited 100 people to develop one day of learning. The 100 Days of Learning edition facilitated distributed events globally for one year. The days of learning took place in local communities, promising a tangible impact by sharing ideas in their own environment. We also closely worked with the Design Academy Eindhoven. Based on our learning from the *Age of Wonderland* process, we developed a module for masters students of Social Design in which students started their research process with a basic question: 'what do we need to learn in order to make the world a better place?' During weekly iterations, we supported students to make their ideas tangible by disseminating what those key moments of insight actually taught them and how to translate this into transferrable lessons to others.

Rather than focusing upon a specific predetermined outcome, the main premise for these two models was to gain new insights by fostering non-trivial interactions between people of various backgrounds and disciplines. The main outcomes of projects devised by our *Age of Wonderland Fellows*[13] included prototypes made in collaboration with local communities, cultural institutions and companies. Through this process, artists became catalysts, bringing questions to the table that normally are not possible within a business environment, challenging companies and institutions, paving the way for social issues and often identifying previously unarticulated urgencies. Some of the artists involved stressed the importance of the local understanding and use – or misuse – of traditional knowledge without analysing its implicit value for contemporary society. Rather than searching for a continuous process of innovation, the practice of the designers was based on the analysis of a very practical local context, its heritage, the popular wisdom or tradition in their region.

Alongside personal interaction, which was a main element in the collaborations we facilitated, the creation of a digital space for sharing knowledge between peers and others, turned out to have a stimulating effect for people to share their findings and experiences. New insights, reports and prototypes were actively shared on existing platforms such as GitHub,[14] Medium[15] or *Age of Wonderland* website.[16]

Collaboration and Experimentation

We realise that a crucial element of both approaches was the opportunity for participants to have face-to-face meetings to get to know more about each other's ideas and processes and to create space for them to connect on a personal level. Baltan added significant value to the process by connecting people whom without our intervention would not have been able to meet. As such, a dedicated space for ideas to emerge and new sparks to ignite was created. This flexible approach towards collaboration and experimentation became a key ingredient of both programs. Both methods consisted of an agile and collaborative approach, informed by a non-hierarchal structure towards research-creation and collaboration. By encouraging this 'enabling environment' to foster informal meetings, leverage was created for new encounters and unexpected collaborations to take off. We coined these chance meetings *organised serendipity*.

During both trajectories, managing people's expectations was of high importance and drove the success of the projects and the overall dynamics of the collaborations we supported. It was important that participants involved actively shared their learning experiences in an open and informal way. Sharing their successes and particularly their failures in a transparent way enabled everyone to reflect upon where the key learnings lay and how they could be applied in their own practices, organisations and contexts. These moments became a key anchor point in our process.

During *Hack the Body*, we facilitated dedicated partner meetings two times a year. The main premise for these meetings was to keep everyone up-to-date, sharing information within a group of trusted peers and partners. The frequency of these meetings was determined by Baltan; however, the development of the process and the content (ideas, technology, research) was steered by each participant individually. Baltan's role was to facilitate these moments by offering a context and a location. We encouraged each partner to become intrinsically motivated and to take ownership of their own development trajectories and duration of the developments. Baltan then leveraged these processes and facilitated new opportunities to realise the expanded potential of the ideas and projects that emerged. An example of this was the development of the WEAR sustain[17] challenge, an EU-funded trajectory that supports creative collaboration for next generation wearable technologies which Baltan co-hosted at our lab.

Age of Wonderland operated like a pressure cooker, contracting the participants' trajectories into a very specific time, location and context, framed by a residency. The emphasis was put on the presentation of outcomes that emerged from collaborations during the timescale of the residency. To conclude their time with us, participants' projects were presented during the Dutch Design Week,[18] which provided a key moment for a spotlight to be put on their projects where the outside world could witness the results.

Context, Roles

Both models developed projects with creatives from a large variety of ideas and backgrounds. We connected participants to (local) organizations and individuals, in order to collectively explore the possibilities with the use of technology. This crossover mentality, in which learning from each other is at the core, is a key element for finding alternative roadmaps for the future.

By inviting international fellows to the Netherlands for *Age of Wonderland* and framing their role as 'experience experts' we ensured that both the public and the network of professionals were equally inspired. Traditionally, Westerners have imposed their knowledge and expertise on the Other as 'missionaries.' Our back-to-front approach generated sincere interest in setting up new kinds of collaborations. On a global scale, we witnessed that people were eager to join the *Age of Wonderland* program. A buzz was created about the program, allowing people to gain new experiences by joining this artists-residency trajectory.

Our definition of artists and designers is very broad. For example, a social entrepreneur from Tanzania was interested in deploying creative strategies for a (low tech) bio-fertilizer. He was invited for the 2015 edition to develop his project on sustainability in food waste. *Age of Wonderland* likes to bring different domains together to explore how these domains become more intrinsically connected. We like to experiment with how knowledge can arise by connecting different values and methods, rather than focus solely on the generation of new knowledge. It is this intrinsic connection that we believe is important within our world today.

Conclusion

There is much more to say about how these models could function as a catalyst for collaboration and innovation. However, my final takeaway is that it is important to understand the ambiguity and complexity of developing novel ways to foreground mutual inspiration and freedom of experimentation.

The ways in which to approach and tackle challenges and achieve the desired ambitions are not always evident. There is no 'one-size fits all' solution. The strategies Baltan employs do not propose a straightforward solution to solve society's *wicked problems*.[19] Working collaboratively and connecting artists, designers, technologists and social entrepreneurs can creates a spark for creativity and new ideas but also requires a strong role as a facilitator. The process demands clarity in how you communicate your expectations between those involved, and to support the alignment of ideas in order to create shared ownership amongst stakeholders. When working with people from different contexts and backgrounds, this helps make the collaborations successful.

We at Baltan continue to aspire to critically reflect upon the implications, promises and pitfalls of pressing issues emerging within our technological society and interrogate the role that people (and creatives) play in its evolution. Our ambition is to raise awareness of alternative models of practice, open up new ways of thinking and develop ideas robust enough to make a meaningful contribution to societal challenges.

Notes

1 Baltan Laboratories: http://baltanlaboratories.org
2 A Baltan topic that connects partners from the cultural and corporate sector to develop projects around biometric data, identity and the human body: http://hackthebody.nl
3 http://ageofwonderland.nl
4 E.E. G. Kiss is a project by artists Herman Maat & Karen Lancel: http://www.lancelmaat.nl/work/e.e.g-kiss/

5 The E.E.G. Kiss software is accessible via GitHub: https://github.com/baltanlaboratories/eeg-kiss-ofapps
6 Social innovation: https://en.wikipedia.org/wiki/Social_innovation
7 Wicked problems are problems that are difficult or impossible to solve because of incomplete, contradictory and changing requirements that are often difficult to recognize: https://www.interaction-design.org/literature/topics/wicked-problems
8 http://ageofwonderland.nl/retrospect/2014
9 http://ageofwonderland.nl/retrospect/2015
10 https://medium.com/@olliga/silence-for-a-change-a-report-about-the-silent-journey-in-kyrgyzstan-20-26-august-2017-3086b5b4b03d
11 http://ageofwonderland.nl/retrospect/2017
12 http://ageofwonderland.nl/fellows/2016/ali-eslami
13 Age of Wonderland Fellow: Our residency guests from Africa, Asia and Latin-America.
14 GitHub - A platform for software developers: https://github.com
15 Medium is a social media platform for sharing (personal) stories: https://medium.com/100-days-of-learning
16 *Age of Wonderland* is a program initiated by Baltan and Hivos to boost global social innovation: http://www.ageofwonderland.nl/
17 WEAR sustain: https://wearsustain.eu
18 Dutch Design week is an international design event that takes place in Eindhoven: http://www.ddw.nl
19 See Note 7.

7

PARTICIPATING IN THE VISCOUS POROSITY OF MAKERSPACES AND FAB LABS

A Participatory Art Perspective

Alexia Mellor

Introduction

In September 2016, I began a placement with Fab Lab Barcelona's European citizen sensing project, Making Sense. Making Sense engaged community groups in Barcelona, Pristina and Amsterdam in addressing local environmental problems by using digital sensors to collect data and then take action. The democratisation of data that Making Sense attempts complements the democratisation of making that fab labs seek. Both reimagine relationships between citizens and governance by applying a 'hacking' ethos to existing models, and both face accessibility issues. In this case, 'hacking' denotes a creative process of repurposing to meet one's need.[1] Technology is a tool to make systems more accessible and aligned with one's needs; however, democratisation equally requires cultivating a context and mindset of active citizenry within communities. Therefore, my collaboration with Making Sense explored how participatory art practices might engage participants to creatively investigate issues of concern while informing democratisation processes within the framework of making culture.

In this chapter, I detail a case study of two Making Sense pilots in Barcelona. The first involved working with community members wanting to learn about sensor technology, while the second engaged a community impacted by high noise levels. While Making Sense is not a participatory artwork, I use the lens of a participatory artist to discuss how the pilots reflect the 'viscous porosity'[2] needed for fab labs to truly democratise making and support active citizenry. This research concludes that participatory art practices can support viscous porosity by bringing non-makers into making culture (including making spaces and fab labs) and by taking making culture out into communities.

Participation

A 'participatory turn' has existed since the 1960s in which various disciplines and professions, ranging from planning to the natural sciences, have sought varying degrees of public input. The current trends of democratising and making transparent different systems that include making, governance and sustainability represent a shift from seeing people as 'participants' contributing to existing agendas, to 'active citizens' critically shaping new worldviews. The Maker Movement and civic tech emerge from this mode of thinking. Civic tech describes technologies designed to make links between private and public sectors and can roughly be divided into two categories: making government transparent and supporting community action.[3] Examples include mobile apps facilitating citizen engagement such as reporting potholes or supporting new forms of voting. While these technologies are focused on democratisation, a digital divide[4] limits accessibility.[5] Civic tech needs entry points to make participants aware of their potential relevance and to give them the confidence to use the technologies effectively.

Fab labs, and the Maker Movement more generally, are critiqued for failing to recognise the socio-political components of maker culture, and the different barriers to accessibility they present, including accessing the physical making space, lack of mentorship and skill sharing, and not addressing different disabilities.[6] Others argue that making is a hobby reserved for the privileged.[7] At the same time, fab labs have the potential to function as living labs[8] in which community members can gather to experiment with technology and making to address local issues. Consequently, I wanted to investigate how my perspective as an artist specialising in participatory strategies might play a role in supporting the development of this context in which participants could critically engage with technologies and share their skill sets and situated knowledge.[9] When defining 'participatory art,' I draw upon art curator Maria Lind's definition of participation as, 'the creation of a context in which participants can take part in something that someone else has created but where there are, nevertheless, opportunities to have an impact,'[10] and sociologist Nico Carpentier's characterisation of participation as a fluid, invitational practice in which individuals can enact power.[11] Together, these definitions support my understanding of participatory art as a collaborative practice involving the creation of a context or platform from which participants can discover and express their agency. During my placement, I explored how Making Sense, through Fab Lab Barcelona, was developing a context to support individuals in applying making and hacking practices to take action on issues of concern.

Understanding Participatory Art

To consider how I use participatory art to create contexts for critical engagement, I reference my ongoing work, *Berwick-Barcelona (BWK-BCN)* (2016–present). *BWK-BCN* is a speculative work that engages participants in collaboratively designing a fictional town formed by merging a market town on the Scottish

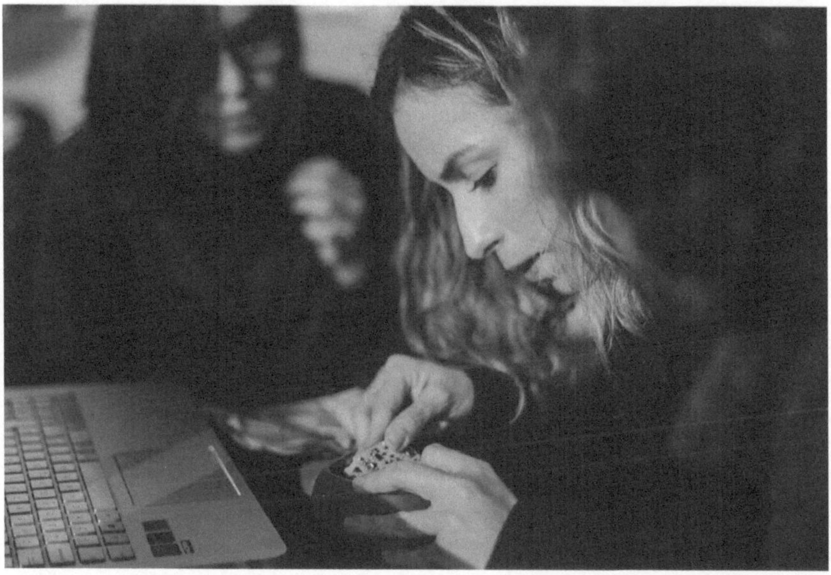

FIGURE 7.1 Smart Citizen Kit. Photo by Making Sense, Gui Seiz.

borders with Barcelona as an adaptation to climate change. Using methods such as working sessions,[12] a citizen science beach walk and a toolkit, participants investigate the relationship between climate and culture, interrogate 'native' vs. 'invasive' species, and uncover their core values that inform local responses to the global issue of climate change. The work remains open-ended and therefore functions as a framework supporting ongoing experimentation, creative thinking and dialogue, as opposed to problem-solving or resulting in a final object. While the work was originally developed in Berwick-upon-Tweed (UK) and involved individuals from various community groups, the resulting toolkit allows people to take part globally, contributing to the project's ongoing nature.

The roles I play in *BWK-BCN* are comparable to Manzini's description of designers as facilitators, triggers and co-design team members that 'mak(e) things happen.'[13] This requires knowing how to engage difficult conversations with differing perspectives and to assist individuals in recognising their tacit knowledge, all while establishing settings to suspend disbelief, imagine different possibilities and to 'hack' current systems.

Making Sense: A Case Study

The Barcelona lab, housed at Institute for Advanced Architecture of Catalonia (IaaC), is a highly active and dynamic member of the international fab lab network.[14] It supports a portfolio of practices, including the international

Fab Academy programme, open source beehives, the future of textiles and the Fab City global initiative that looks to create new urban models of 'locally productive and globally connected self-sufficient cities.'[15] Located in the Poblenou neighbourhood of Barcelona, the city's former industrial area, the fab lab has worked closely with the local municipality to revive Poblenou as a maker district.[16] With its wide-reaching initiatives and collaborations with the public sector and private companies such as IKEA, Fab Lab Barcelona is quickly establishing itself as a major hub within a making ecosystem that links its local activity with global sustainability agendas. As part of this ecosystem, Making Sense[17] (2015–2017) was a citizen sensing project funded by the European Union's Horizon 2020 Collective Awareness Platforms for Sustainability and Social Innovation (CAPS) programme. Compared to citizen science projects in which project organisers establish research agendas, Making Sense participants identified objectives and used the open source/open hardware Smart Citizen Kit (SCK) sensors developed by Fab Lab Barcelona to collect environmental data and take local action.[18] An example of civic tech, Making Sense uses SCK sensors to support citizens in gathering the necessary information to better understand factors impacting their environments and to decide on a course of action. As an artist collaborator, my interest was in how technology could be used as part of a wider context to catalyse ongoing active citizenry and how the data collected could be harnessed to critically look at relationships between human and natural systems and engagement with public space.

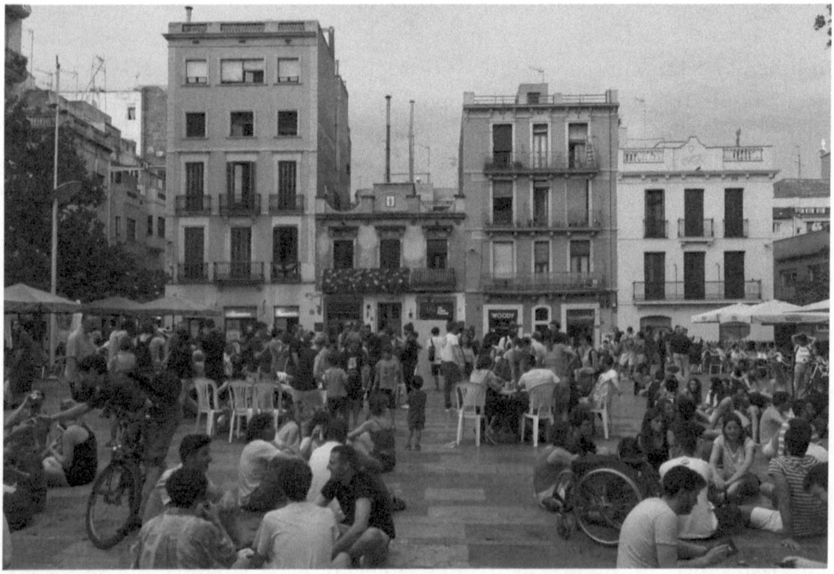

FIGURE 7.2 Plaça del Sol, Barcelona. Photo by Saskia Coulson.

In the sections below, I describe two of the Barcelona pilots. Viewed together, they reflect the dual aspects of viscous porosity: how fab labs might bring together non-makers with makers, and how fab labs might extend their reach and the making ethos into communities.

Description of Pilots

The first pilot consisted of approximately 25 citizens from across Barcelona who attended a public information session held at the fab lab. While some members had identified particular environmental issues in their neighbourhoods, the majority were curious about the sensors and citizen sensing. They represented a range of ages and levels of familiarity with digital technologies. Over the course of several weeks, we trained the participants to use the sensors to monitor the data and to develop a public intervention that would serve as a platform to share their learnings while encouraging new participants to take part. During the trainings, we made the entire process transparent and included the participants in identifying technical issues, as well as collecting their feedback on how to improve the onboarding process. They gained the skills and confidence to assist future community groups in using the technology to address their own issues, and they were invited to work with the group in the second pilot – a community of residents living in Plaça del Sol, a busy square in the popular neighbourhood of Gracià. The Plaça del Sol residents wished to use sensors to address an acute issue of noise pollution. As a large and densely populated city and popular tourist destination, Barcelona faces significant noise pollution. Plaça del Sol is a favourite square for locals and tourists alike, and the trend of late-night drinking is a problematic issue for the residents.

To address the noise problems in Plaça del Sol, the second pilot took place exclusively in a co-working space located near the residents' homes. These participants were less interested in the sensors or making culture, and more concerned with how technology might provide important data to illustrate the severity of their noise problem. The co-working space became our hub, regularly bringing together community members, neighbourhood associations and experts to discuss the noise issue. The residents worked with the Making Sense team to develop a strategy for monitoring noise levels in the exterior and interior of their apartments, ensuring that data was collected from different storeys of the buildings in order to gain a clear picture of the situation. This required collaboration between neighbours, with some residents serving as representatives of others who were less interested or unable to attend the meetings, but who were willing to host sensors in their home. Together, we mapped the systemic relationships between noise and businesses, rubbish collection and socialising patterns in Plaça del Sol.

Developing the pilots' engagement strategy, I worked alongside the project's core team to organise workshops, facilitate activities and assist participants in

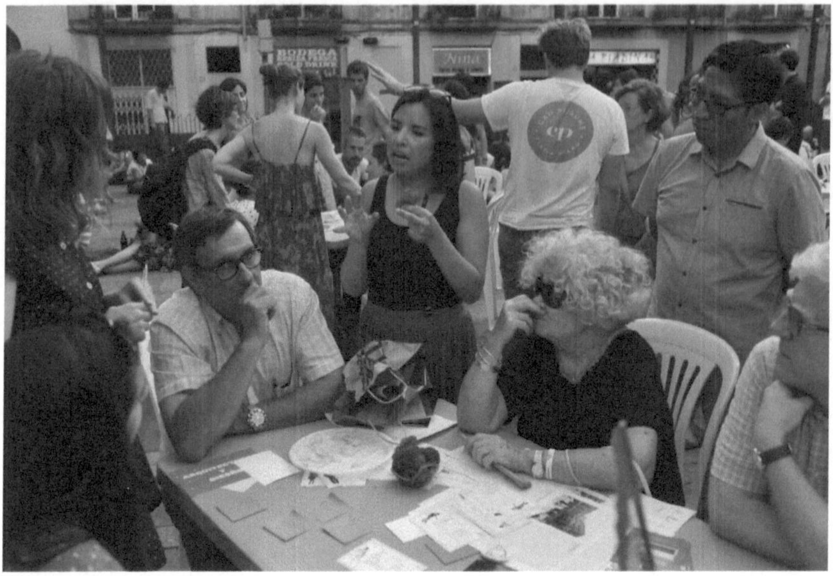

FIGURE 7.3 Plaça del Sol Community Roundtable. Photo by Saskia Coulson.

clarifying their concerns, while experimenting with scenario building and narrative development exercises. I collaborated with participants in the first group to develop an installation using an SCK and LED strip to visualise noise levels in public space and to initiate conversations around the effects of noise pollution. Using my artistic background, I challenged participants to consider the installation's intention, and the details that would help bring passers-by to engage with the piece, making the data more relevant to them. In the second pilot, we focused on creating space for discussing the issue, deciding where to place the sensors, and comparing official city data with the neighbours' collected data. The pilot culminated with an event and public installation using cut-outs of the neighbours' silhouettes made in the fab lab to bring awareness to the plaza as a residential area. During the outdoor event, a series of roundtables with Plaça del Sol residents and other curious onlookers invited discussions on noise pollution from multiple perspectives, including activities that could replace existing ones in the square, how new materials and architecture could change the experience of noise, and an open table where individuals could suggest additional topics.

Discussion of Outcomes

Science and Technology Studies researchers Neumann and Star describe how infrastructure becomes visible only when it fails,[19] and the Making Sense pilots provided important learning regarding tensions and limitations to the fab lab infrastructure. Given the first pilot's focus on technical training, the majority of these workshops took place at the fab lab. The layout of the lab embedded within IaaC is not intended to meet the needs of those entering from outside. This hindered integrating the Making Sense participants into the fab lab community. To enter the space, visitors must ring a buzzer operated by the IaaC office team, and therefore access is limited outside office hours. The fab lab also lacks a dedicated space where groups can meet, discuss and work together. As a result, we were often forced to hold workshops in the unheated open area of the lab, which was regularly disturbed by the loud machines or students working on projects. While seemingly insignificant, from my experience of establishing appropriate contexts for participation, the lack of a welcoming space significantly impacts how participants engage and their willingness to experiment. Uncomfortable in the open space, participants requested a different site for the workshops.

Despite many participants being interested in learning to use the lab's machinery, we could only host a one-day introductory workshop due to limited availability of machines and technicians, particularly during busy end-of-semester periods. Given Fab Lab Barcelona's vision of better integrating into the surrounding community, there is a real need to address accessibility to the space, and lay the necessary groundwork for an ongoing, developing relationship with the neighbourhood. This is where participatory artists add value. My practice employs the methodology of infrastructuring, borrowed from Participatory Design to describe an open-ended and iterative making process that remains open for future participants.[20] Infrastructuring involves understanding particular contexts, and developing strategies for continuous engagement that responds to participants' changing needs.[21] It entails creating entry points for new participants to join at any point in the work. The first pilot revealed the need for a constant onboarding process with the technology, in which technical experts were required to regularly troubleshoot, assist participants in getting the sensors online and make them feel comfortable and confident. Therefore, regular, strategic engagement on all levels must become an embedded practice for fab labs in order for making culture to be truly democratic.

For the second pilot, the Making Sense team met the Plaça del Sol residents in their community, and helped troubleshoot technology issues onsite. This created a different relationship to fab labs in which making culture was extended beyond the lab into the co-working space and people's homes. It should be noted, however, that despite the distributed nature of this pilot, the central meetings at the co-working space were invaluable for establishing a sense of camaraderie and shared purpose. Like the first pilot, much time was spent on the onboarding of technology. Sharing this experience alongside their neighbours proved to be important to participants feeling ownership of the technology and confident using it. Participants in both pilots cited the experience of being part of a community with shared interests, and feeling empowered to take action, as key outcomes. They continue to feel encouraged by the project's success, which has garnered media attention and has led to the Barcelona City Council making improvements, including changing cleaning times to 11pm from 3am as a deterrent to people staying late in the square, and avoiding late night disruptions to the neighbours.[22]

Both pilots reflect Lindström and Ståhl's concept of 'becoming the response-able stakeholder,[23] a concept referring to creating the appropriate context for individuals to discover their own entanglements with an issue of concern,[24] followed by exploring their ability to respond. I have applied this approach to my practice in order to assist in articulating issues of concern and identifying the socio-material assemblies that form them.[25] While all of the residents were distressed by the noise in their square, they each had their own narrative that we explored through the workshops. Equally, participants in the first pilot had varying reasons for wanting to learn about the sensors, and for considering how technology could empower citizens with the necessary tools to capture data and make a case for change, therefore facilitating new ways of engaging with local government. By creating multiple entry points and by supporting individuals to find personal connections, Making Sense began the process of developing a bottom-up context for exploration and experimentation. Participatory art methods played an important role in supporting this context by teasing out personal narratives to gain a fuller picture of the complexity of the issues of concern, including considering noise as a symptom of the larger question of how we use public space. Such an approach counters solution-focused efforts, and instead supports broader critical engagement to foster understanding of issues from multiple vantage points.

Conclusion

Making Sense has had a tremendous impact on my thinking around the potential role that participatory artists can play in the viscous porosity of fab labs. With challenge-led research initiatives into wicked problems calling for wider participation, a cultural shift is needed regarding how we think about who contributes to shaping agendas and how we can redefine systems. This requires establishing a broader definition of making that focuses less on the technology itself, and more on individuals' rights and responsibilities to take an active role in creating the world they want to see. Fab labs must take an ecosystems approach, gaining better understandings of what communities want and expect from making spaces, while also facilitating community members to imagine different possibilities. Fab labs offer an ideal space for sharing technical skills and supporting experimentation, but they need assistance to become more widely accessible to diverse groups of people, and to change the perception that they are only places for those interested or skilled in technology. Participatory artists located in fab labs can support the sharing of so-called 'soft skills' and critical questioning that looks beyond the object and asks who is making what, and what drives the making. Participatory art practices using an infrastructuring approach have the unique opportunity to support fab labs in establishing an environment for ongoing and responsive engagement, discovering new ways to frame issues of concern and transforming participants into active citizens and makers who 'hack' and repurpose systems to meet their needs.

Notes

1 Rosner, Daniela K., and Jonathan Bean. 2009. "Learning from IKEA Hacking: "I'm Not One to Decoupage a Tabletop and Call It a Day."" *CHI 2009* pp: 419–422.
2 Feminist philosopher Nancy Tuana refers to "viscous porosity" to describe the blurring of boundaries that occurs when humans and social practices interact with the environment and natural phenomena. The term "viscous" recognises points of resistance when moving back and forth between these physical and symbolic boundaries, and how this offers the potential for new knowledge and understandings. I appropriate the term and apply it to the fab lab context to describe the reflexive relationship between making culture and the wider community; recognising the need for physical porosity and permeability in which maker spaces are made accessible to and informed by the public, and how the difficulties within this process can reveal opportunities for new knowledge. Tuana, Nancy. 2008. "Viscous Porosity: Witnessing Katrina." In *Material Feminisms*, Stacy Alaimo, Susan J. Hekman (eds), 188–213. Bloomington: Indiana University Press.
3 Mayur, Patel et al. 2013. "The Emergence of Civic Tech : Investments in a Growing Field." http://www.knightfoundation.org/media/uploads/publication_pdfs/knight-civic-tech.pdf, 30.
4 The "digital divide" refers to differences in access to digital technologies, as well as differences in how these technologies are used, as a result of social inequalities.
5 Eastin, Matthew S. et al. 2015. "Extending the Digital Divide Conversation: Examining the Knowledge Gap Through Media Expectancies." *Journal of Broadcasting and Electronic Media* 59 (3): 416–437.
6 Alper, Meryl. 2013. "Making Space in the Makerspace: Building a Mixed-Ability Maker Culture." https://teethingontech.files.wordpress.com/2013/03/idc13-workshop_meryl-alper.pdf.
7 Ames, Morgan G. et al. 2014. "Making Cultures: Empowerment, Participation, and Democracy—or Not?" *Proceedings of the Extended Abstracts of the 32nd Annual ACM Conference on Human Factors in Computing Systems – CHI EA '14*, 1087.
8 Björgvinsson, Erling et al. 2010. "Participatory Design and 'Democratizing Innovation.'" *Proceedings of the 11th Biennial Participatory Design Conference on – PDC' 10*, 41.
9 Haraway, Donna J. 1988. "Situated Knowledges: The Science Question in Feminism and the Privilege of Partial Perspective." *Feminist Studies* 14 (3): 575.
10 Lind, Maria. 2007. "Taking the Matter Into Common Hands: On Contemporary Art and Collaborative Practices." In *On Contemporary Art and Collaborative Practices*, J. Billing et al. (eds), 17. London: Black Dog.
11 Carpentier, Nico, 2012. "The Concept of Participation. If They Have Access and Interact, Do They Really Participate?" *Fronteiras – Estudos Midiáticos* 14 (2), 170–172.
12 I use the term working session instead of workshop to emphasise the ongoing and open nature of the work.
13 Manzini, Ezio. 2014. "Making Things Happen: Social Innovation and Design." *Design Issues* 30 (1), 66.
14 *Fab Foundation*, http://www.fabfoundation.org/ [Accessed March 19, 2018].
15 *Fab City*, http://fab.city/ [Accessed March 5, 2018].

16 *Barcelona Digital City*, http://ajuntament.barcelona.cat/digital/en/digital-innovation/make-in-bcn/maker-district [Accessed March 5, 2018].
17 *Making Sense*, http://making-sense.eu/ [Accessed March 5, 2018]
18 Balestrini, Mara et al. 2015. "From Participatory Sensing to Making Sense." In *Environmental Infrastructures and Platforms 2015 – Infrastructures and Platforms for Environmental Crowd Sensing and Big Data Co-Located with the European Citizen Science Association General Assembly 2015 (ECSA GA' 2015).*
19 Neumann, Laura J. and Susan Leigh Star. 1996. "Making Infrastructure: The Dream of a Common Language." In *PDC '96 Proceedings of the Participatory Design Conference*, 231–240.
20 Karasti, Helena. 2014. "Infrastructuring in Participatory Design." In *Proceedings of the 13th Participatory Design Conference on Research Papers – PDC '14*, 141–150.
21 Björgvinsson, Erling et al. 2010. "Participatory Design and 'Democratizing Innovation.'" Proceedings of the 11th Biennial Participatory Design Conference on – PDC' 10, 41.
22 Fildes, Nic. 2017. "Grassroots Groups Use 'Internet of Things' Data to Tackle Damp and Noise: Data-Collecting Gadgets Help People Hold Authorities to Account." *Financial Times*, September 26, 2017. https://www.ft.com/content/2a81afcc-7204-11e7-93ff-99f383b09ff9 [Accessed March 19, 2018].
23 Lindström, Kristina, and Åsa Ståhl. 2016. "Becoming Response-Able Stakeholders: Participatory Design in Times of Uncertainties." In *The 14th Participatory Design Conference*, 41–44.
24 Marres, Noortje. 2007. "The Issues Deserve More Credit: Pragmatist Contributions to the Study of Public Involvement in Controversy." *Social Studies of Science* 37 (5), 759–780.
25 Björgvinsson, Erling et al. n.d. "Design Things and Design Thinking: Contemporary Participatory Design Challenges." *Design Issues* 28 (3): 101–116.

+ CONTEXT III

Engaged Communities

8

IYAPO REPOSITORY

Constructing and Archiving Alternate Futures

Ayodamola Tanimowo Okunseinde

Introduction

Iyapo Repository is a resource library that houses a collection of digital and physical artifacts created to affirm and project the future of people of African descent. The collection is managed and developed through a series of workshops where participants become archivists of a future they envision. The resource library holds workshops in which participants sketch out and rapid prototype future artifacts in domains such as food, music, politics and fashion. These sketches constitute the collection of manuscripts. The repository then works to bring a select few of these artifacts to life. They become technologically functional while staying true to the participants' original blueprints. Alongside the art and artifacts collection, Iyapo Repository also hosts manuscripts, films and rare books.

Iyapo Repository has presented workshops at Eyebeam, Brooklyn Museum and The Museum of Contemporary African Diasporic Art (MoCADA). Through a residency with The Laundromat Project, the repository worked with The Bed Stuy Museum of African Art to build a curriculum around future design thinking and prototyping methods that promotes self-determination in technology. With Carnegie Mellon University, the repository introduced a workshop component in which participants use virtual reality to sketch and engineer future artifacts. The repository has been exhibited at numerous institutions including The Laundromat Project, Macalester College, Recess Assembly, and the 11th Shanghai Biennale. Iyapo Repository was developed by Salome Asega and Ayodamola Okunseinde as part of their 2015–2016 Eyebeam Project residency.

Genesis

September 2014, Union Square, New York, NY.

I was afraid, and imagined news reports detailing my death, the call to my parents. What terrified me most was the inevitability of the encounter. Recent shootings of Black bodies, protests, police brutality. Lives extinguished. These thoughts and myriad of others, injustices, violence, memories of personal attacks on my body, slurs, the denial of my personhood… swirled through my mind. I was a graduate student, I was from an upper middle-class family, parents educated, well-traveled, and yet it seemed my Blackness still led me here. Quaking in fear on the ground assaulted by this officer for no apparent reason, a possible statistic.

I began by contemplating the fear that no matter what I did, who I was, I could never escape my Blackness and these violent consequences. This had not been the case in other countries I lived in, Nigeria, Oman, Holland, where it seemed my humanity came before my Blackness. In America, my humanity is and has always been flattened to a thin monolithic black line devoid of nuance, squeezed of culture, and history. It was that flattening that allowed the officer to treat me recklessly. I felt compelled to make a statement, to expand this flatness of Blackness through my work. The process of reconciling these thoughts into a cohesive manifestation consumed me for several months. I resolved on the following list as a means of catharsis:

- Work should be interactive, performative and live in the world, thus maximizing the potential for interaction and serving the public good.
- Work should be multi-dimensional projections in which African artifacts, food, language and cultural elements are highly represented.
- Work should be healing and affirming of the diversity of Blackness and the people of African descent.
- Work should exist in technologically speculative and virtual spaces.
- Work should be primarily driven by narratives and storytelling methods.

The Rift: An Afronaut's Journey

In adhering to this list, my graduate thesis work, *The Rift: An Afronaut's Journey*,[1] was born with the aim of forcing dialog and contemplation on the differences between the speculated future, past, and present of people of African descent.

The narrative that developed revolved around future archaeological objects found by the Afronaut. These objects (Suit, Helmet, Food Artifact, Pollution Artifact, and Communication Artifact) are a fusion of current and future Yoruba artifacts. The Afronaut is simultaneously human and alien, of the present and of the future. He is other, yet uncannily familiar.

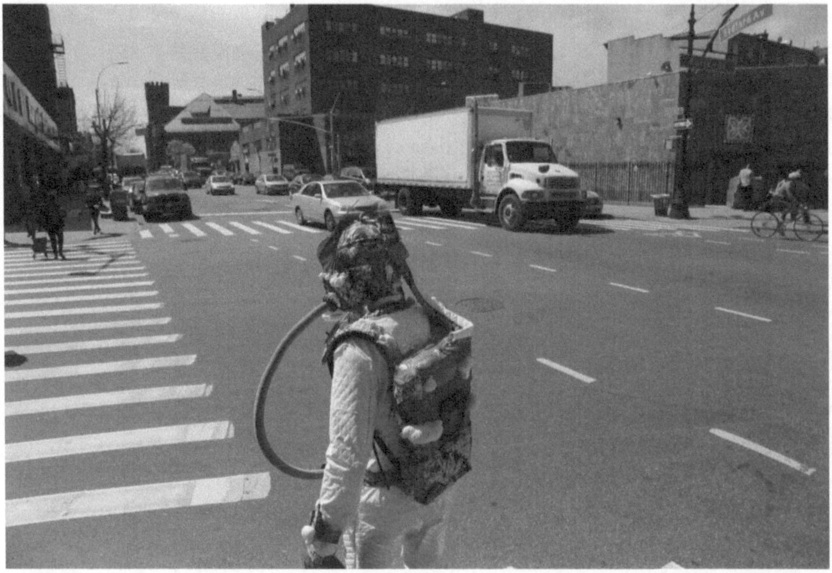

FIGURE 8.1 The Afronaut in Bedford-Stuyvesant, NY. © Ayodamola Okunseinde. Photo by Stephanie Rose Burgess.

As one studies artifacts of the past to understand past cultures, likewise one can study artifacts of the future to understand future cultures. This Archeo-future[2] concept would become the key narrative framework for the Iyapo Repository. To understand the emergence of the Iyapo Repository, it is important to understand the study of the Afronaut's artifacts as an investigation into his culture.

Whereas *The Rift: An Afronaut's Journey* was an expression of my frustration of America's flattening of my Blackness, I considered it critical to expand the project to allow other voices to express their hopes, fears and perspectives as members of the African diaspora in the context of 21st century America. The opportunity came in 2015 when I applied for residency at Eyebeam, a critical institution in New York that focuses on technology and society. The original proposal for Eyebeam, entitled *Prophecy: Artifacts of African Futures* envisioned the following:

1. Vision: where ideas were generated communally.
2. Repository: where generated ideas were archived.
3. Projection: where the archives would be manifested.
4. Exhibition: where the work would be presented.

Collaboration

The project would need to develop methods to ensure community partners would be heard, considered and in no way feel they were being exploited. At the time,

I hadn't made participatory artworks and realized I would need a collaborator with experience implementing participatory projects. I knew Salome Asega, a fellow alumnus of the Design and Technology program at Parsons the New School of Design, and was impressed with two of her works in particular: *Level Up: The Real Harlem Shake*, where she worked with dancers and interactive technologies to teach audience and participants about the Harlem Shake.[3] In *Crown Heights Mic*, Salome created a collaboratively broadcast network built by neighborhood stakeholders and residents. Salome's expertise with community-driven workshops and artworks was thus invaluable to the trajectory and success of our project.

Eyebeam

We spent the first month of our residency engaged in creative research. We plastered the walls with inspirational images, projected related movies and music videos, engaged in vigorous discourse about the nature of community-driven works and discussed the structure of the physical, virtual and social spaces the repository would occupy. The three primary considerations for our work at Eyebeam were the development of the narrative, articulating the functionality of the work and contemplating the aesthetic and formal elements of the works that would emerge.

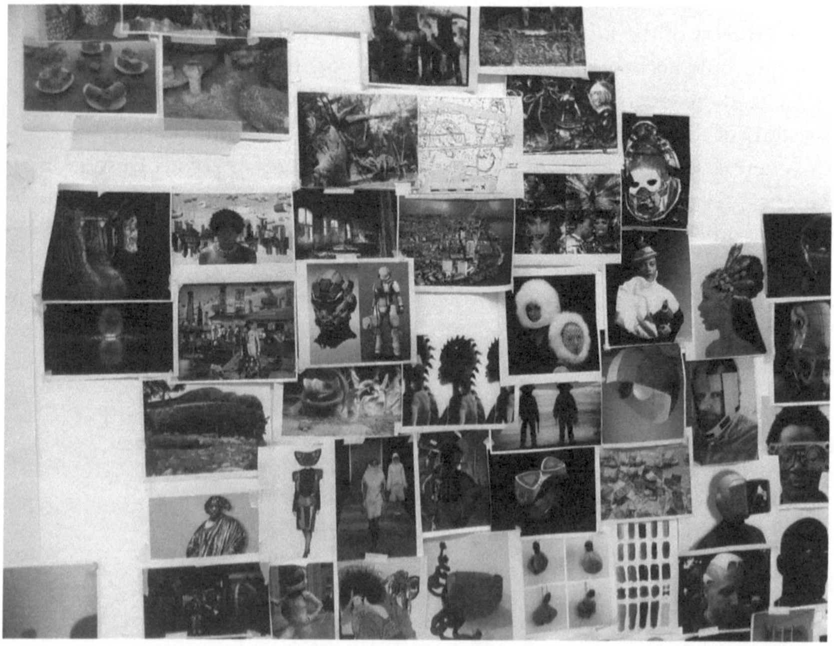

FIGURE 8.2 Eyebeam studio walls with research material. © Ayodamola Okunseinde and Salome Asega.

As designers, we are regularly confronted by constraints. As artists working with technology, we regard constraints as creative challenges and not restrictions. The constraints of time, finances or technology can thus be leveraged as tools that help define the final outcome of a project.

Function

Traditional African art is a functional and necessary part of everyday life that would be impossible to comprehend out of context. As the *Iyapo Repository* concept emerged, we wanted to ensure that the final artifacts produced were similarly utilitarian. We insisted that the artifacts and our community-driven workshops be a portal to manipulate and control the future via audience interaction. The original title, *Prophecy*, sprung from these ideas. Soon, however, we renamed our project *Iyapo Repository*, which is a reference to Octavia Butler's protagonist in her *Lilith's Brood*[4] trilogy. The new title, being less literal, allowed us more latitude in formulating the project's narrative.

Narrative

The narrative revolves around a research center that exists in a nondescript future and collects African artifacts from this future's past. We allowed for the assumption that both the past and the future could exist in the same dimensional space. We wanted participants to imagine all possibilities and to approach their ideas without limitation.[5]

By collecting and researching these archaeological artifacts, archivists of the future museum gain an understanding of the lives and cultures of their African ancestors. The notion being that as we understand cultures of our past through their artifacts, we can understand cultures of our future through the artifacts housed in this future museum, and unearthed by future archivists (our workshop participants). It is important to note that we envisioned the repository not only as a museum but also as a cultural information nerve center. We wanted to ensure that we projected a positive representation and did not reproduce tropes of ethnographic museum misrepresentations, such as the haphazard collections of The Trocadero Ethnography Museum.[6] The question of what a museum of the future would look like was thus driven by what sorts of experiences we would want visitors to have. We visited several museums, including the Schomburg Center for Research in Black Culture, where we found great inspiration for our center's structure. We modeled the Iyapo Repository loosely after the divisions of the Schomburg Center:

- *Arts and Artifacts Division:* contains artifacts that are three dimensional in form.
- *Manuscript Division:* holds the original manuscripts and sketches produced from our workshops.

- *Moving Image Division:* reserved for photography and films made of the artifacts in use and in context.
- *Rare Books Division:* contains zines and books from all over the world in an online portal or in our Dead Drop Library,[7] a series of mounted USB ports containing curated digital material.

Workshop

The 90-minute design thinking, community-driven workshops were the lattice for unifying and bringing to life the narrative and functional aspects of the Iyapo Repository project. They also served as the means to collect the research material that would subsequently be made into artifacts. We worked to develop a method that was not only instrumental in collecting material but also allowed for discussion and debate about the significance of the work created. The method would also need to blur the lines between reality and the project narrative to keep in line with a Speculative Design[8] practice. After researching potential methods, we loosely integrated elements of Situation Lab's *The Thing From The Future,*[9] a collaborative game in which payers describe objects from alternative futures.

The workshops began with an introduction to the *Iyapo Repository* history and narrative structures via multimedia slides. Participants were then divided into groups consisting of four to six individuals. Groups became archivists of the future repository, then were allotted cards to gets them to think of the future in different domains (i.e.: politics, fashion, food, health). Their task was to uncover and document archeological artifacts from this imagined future's past.

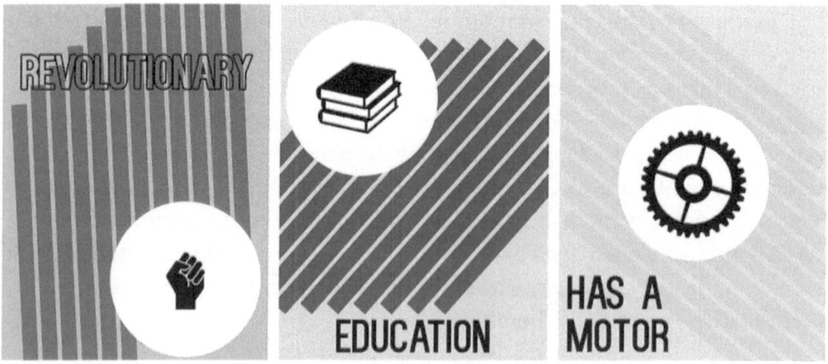

FIGURE 8.3 Iyapo Repository workshop cards. © Ayodamola Okunseinde and Salome Asega.

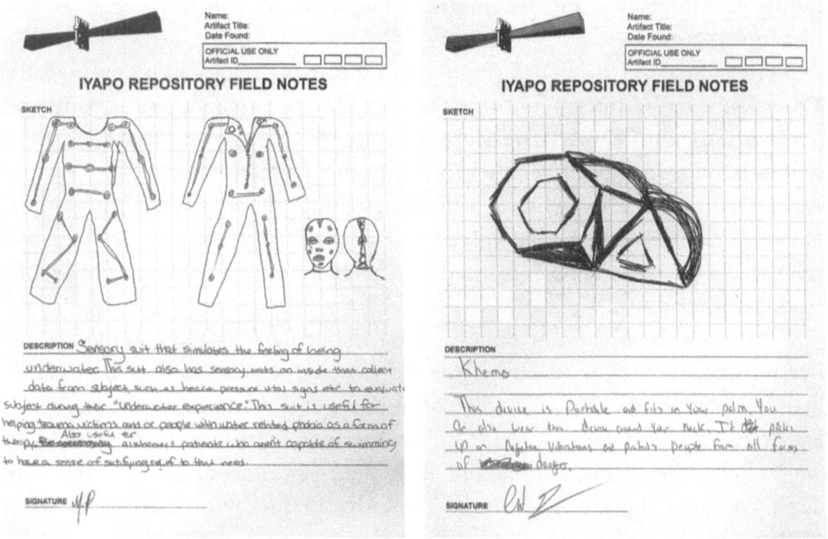

FIGURE 8.4 Manuscripts of Artifact_012 (left) and Artifact_025 (right). © Ayodamola Okunseinde and Salome Asega.

In the above example, an archivist (a workshop group) would have to envision a revolutionary educational tool that somehow incorporates a motor. After ideation, they would sketch out this future artifact on a repository Field Note document (Figure 8.4). In some cases, we provided craft and electronics materials for the participants to roughly prototype their designs. We then asked each group to describe the future artifacts and their social, political or cultural significance. We kept records of these interactions and conversations in order to properly contextualize the artifacts for fabrication.

Fabrication

Artifacts selected for final fabrication were evaluated based on their design practicality, technical feasibility and how closely they adhered to our conceptual narrative. We utilized laser cutters and 3D printers, which we had limited experience using prior to Eyebeam. We used the laser cutters primarily for engraving and cutting acrylic parts for artifacts and manuscript holders. The 3D printers were instrumental in enabling us to rapidly prototype potential forms, then finalize artifacts. In cases in which the manuscript clearly prescribed the form of the artifact, we adhered closely to the archivists' findings. In cases in which the prescriptions were lacking, we took artistic liberty to fill in the details. We interpreted the manuscripts as one might interpret ancient directions, taking care to glean as much from the context around which the primary document was created.

For example, Artifact_025 (Figure 8.5) specifies '... it picks up on Negative Vibrations...'. We interpreted this to mean that the artifact senses dangerous geolocations, thus the artifact alerts the wearer to geolocations where Black bodies have been extrajudicially killed. For this artifact, Khemo,[10] we embedded a GPS unit that syncs with a database and illuminates when the user encounters locations of police-involved shootings of Black bodies. We embedded the parts in a translucent 3D-printed PLA shell that mimics the geometric form from manuscript.

Fabrication involved not only creating the artifacts but also building supporting hardware. For example, the display of the manuscripts necessitated manuscript holders. We created these using wood, aluminum and acrylic in Eyebeam's woodshop. As a result of our experience gained in the woodshop, we later created wood, acrylic and aluminum pedestals to present final artifacts. This gave the display a finished look that further reinforced the museum aesthetic.

With assistance from associates and general exposure to state-of-the-art fabrication methods, by the end of our time at Eyebeam, we had gained extensive knowledge of the possibilities and limits of laser cutters, 3D routers, woodshop, and other prototyping tools and methods. These skills have remained instrumental in the development of the *Iyapo Repository*.

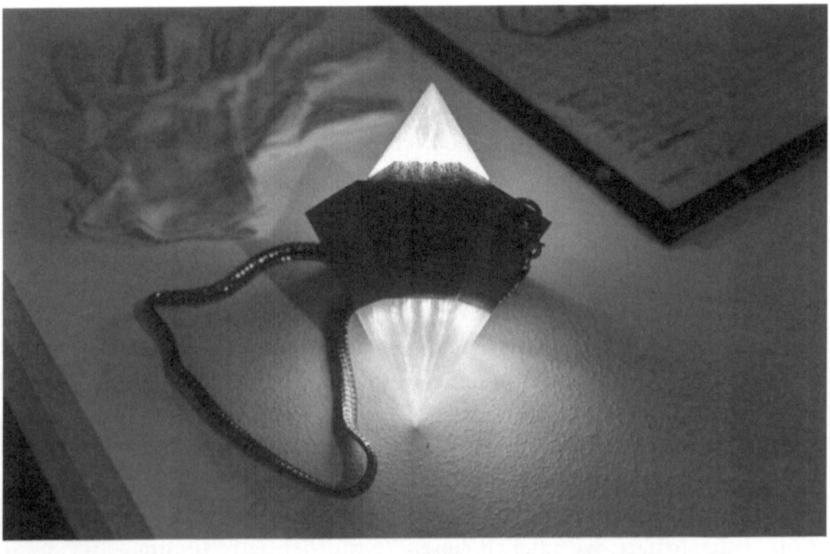

FIGURE 8.5 *Artifact_025.* © Ayodamola Okunseinde and Salome Asega.

Films

To add additional narrative context to the selected final artifacts, we created short films, *Water* and *Mother Radio*. These films demonstrate the artifacts' technological and magical powers in-situ with the narrative. Myself, Salome, and team members Mala Kumar and Mariama Jalloh, in conjunction with photographer Derek Schultz, created two films that evoke a sense of magical surrealism as they transport viewers into the time-space of *Iyapo Repository. Water* (Figure 8.6) depicts the use of Artifact_012 as prescribed by Manuscript_012 (Figure 8.4). We continue to make films for our selected artifacts as *Iyapo Repository* develops.

Exhibition

Beyond technical and research support, Eyebeam provided a method to test our narrative and workshops through their 'Stopwork' program. These Eyebeam critique workshops enabled us to invite our current residency cohort alongside alumni to interact with elements we wanted to test. This context greatly sharpened our project. Additionally, the constant flow of friends and studio visitors at Eyebeam turned out to be crucial in helping us clarify issues of function and presentation for the repository.

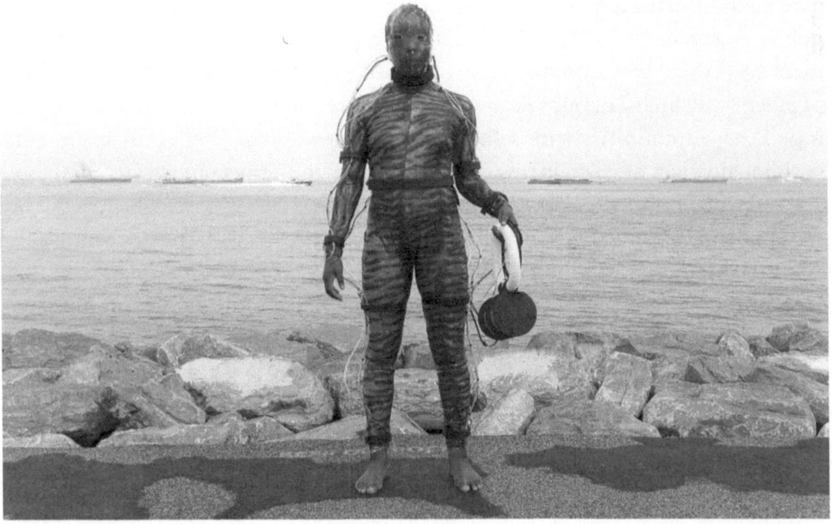

FIGURE 8.6 Still from the film *Water.* © Ayodamola Okunseinde and Salome Asega. Photo by Derek Schultz.

Our residency culminated in an exhibition of the residents' works. This context placed the newly founded museum of the *Iyapo Repository* next to a gamut of creative technologies from biohacking to science fiction and virtual reality. At the close of the residency, we felt we had created robust artworks and a strong narrative foundation. Next on our agenda was to learn from the exhibition of the works in other spaces.

What We Learnt

During an interview explanation of our work to a Taiwanese TV crew, we were asked why we build such 'sad works.' Salome and I were struck by the question. We had never considered our work sad but rather critical and full of hope. With this question, the notion of context came to the foreground. We reasoned that the TV crew, viewing our work outside of the context of American history and current affairs, could only glean from it ominous overtones. This realization re-emphasized the complex layers of the works and was an opportunity to learn how others may perceive our creations. With each iteration and presentation of the *Iyapo Repository*, we continue to learn inspiring lessons. We articulate some key takeaways below.

The Laundromat Project

At the culmination of our residency at Eyebeam, we were accepted as members of The Laundromat Project, another critically important institution for the arts and public engagement in New York City. We participated in numerous development workshops that aimed to hone our skills in interacting with our community of practice. Primary in our takeaway from this residency was the notion of listening to the communities with which we work. It is essential not only to hear the individual voices of participants through their manuscripts but also to synthesize and reflect communal voices as a whole as through the works we collected during our residency. The works are representations of an African diasporic community. Thus, the question of who makes the works (via workshop participation), and by which community the work is meant to be viewed (through exhibition) is a question we are still trying to understand. We know we want the project to be framed with respect to the African Diaspora, but we also want to ensure that the project is inclusive enough to relate to other communities.

Recess Assembly

Recess is an organization that fills the space between gallery and private studio. As artists-in-residence in their Downtown Brooklyn Recess Assembly gallery, we worked with individuals, aged 16–24, who had been convicted of misdemeanor crimes in Brooklyn. Via court mandate, this group participated in a four-week Iyapo Repository workshop during which we asked them to apply their history to our process. Though ultimately successful, we were struck by the challenges the groups faced in envisioning positive futures for themselves. We attributed these challenges to rising from structural inequalities that insidiously curtails the space to dream[5]. The potential for our work to address or shed light on these inequalities brought a greater sense of accountability that we hold close as we continue our work. Recess Assembly showed in a practical sense, how our works and process may function as tools for healing and/or catharsis and underscored the critical nature of our work to reach certain communities.

Macalester College

Our presentation at Law Warschaw Gallery, Macalester College (Saint Paul, Minnesota) in the Fall of 2017 demonstrated the most extensive exhibition of the *Iyapo Repository* to date. The programing included several workshops at multiple locations around Saint Paul, studio visits and engagements with local institutions, as well as casual conversations with students and faculty of the university. We presented over 250 of the repository manuscripts, over 20 artifacts and the repository films. The presence of all the works in concert with the programming gave a sense of gestalt that affirmed the project's narrative, as well as a validation of the artifacts themselves. Additionally, it signaled the potential gains from a scaling up of the project and shed insight as to how we may utilize scale as an integral method to maintain an immersive narrative. The experience allowed us to collect a plethora of new manuscripts and sparked ideas on how the repository may grow.

Conclusion

The *Iyapo Repository* has grown beyond what we envisioned in 2015. We now regularly exhibit nationally and internationally, we have recently added Mala Kumar and Mariama Jalloh as valuable new members of our team. We are currently investigating new methodologies of interaction such as AR/VR and working with partners on new themes of immigration and gentrification.

Perhaps the most exciting project we are working on is conducting our workshop in other countries that have large African Diaspora communities such as Jamaica, Brazil or Cuba. How do individuals of these communities project their identities, how can these communal hopes and fears be manifested in artifacts, and how do these artifacts differ or add to the understanding of the artifacts already collected. Is there a global consensus of how communities of the African Diaspora understand their past and envision their future? Contemplating the output of upcoming workshops and projects fill us with excitement as we anticipate continued learning from the growth of the *Iyapo Repository*.

Notes

1 Okunseinde, Ayodamola Tanimowo. "*The Rift: An Afronaut's Journey.*" 2015, http://digitalarchives.library.newschool.edu/index.php

2 Archeo-future is similar to Guillaume Faye's Archeofuturism only in its phrasing and not concept. The Archeo-future expressed here is more conceptually tied to the works of Laura Forlano. Forlano, Laura. 2013. "Ethnographies from the Future: What Can Ethnographers Learn from Science Fiction and Speculative Design?" *Ethnography Matters*, September 26, 2013. https://ethnographymatters.net/blog/2013/09/26/ethnographies-from-the-future-what-can-ethnographers-learn-from-science-fiction-and-speculative-design/.

3 Harlem Shake refers to a dance originated by Harlem resident Al B. that went viral as a meme in 2012. Gregory, Kia. 2013. "It's a Worldwide Dance Craze, but It's Not the Real Harlem Shake." *The New York Times*, February 28, 2013. https://www.nytimes.com/2013/03/01/nyregion/behind-harlem-shake-craze-a-dance-thats-over-a-decade-old.html.

4 Osherow, Michele. 2000. "The Dawn of a New Lilith: Revisionary Mythmaking in Women's Science Fiction." *NWSA Journal* 12 (1): 68–83.

5 *Imagination as Privilege*, by Allison Freedman Weisberg and Shaun Leonardo, provides deeper insight on how factors such as poverty, lack of resources, access to education and health care foreclose possibilities.

6 Barasch, Moshe. 1998. *Modern Theories of Art 2: From Impressionism to Kandinsky.* New York: New York University Press.

7 Our Dead Drop Library references the dead drop method of espionage communication, as well as fellow Eyebeam Alum Aram Bartholl's *Dead Drops* USB project.

8 Dunne, Anthony, and Fiona Raby. 2013. *Speculative Everything: Design, Fiction, and Social Dreaming.* Cambridge: The MIT Press.

9 Candy, Stuart and Jeff Watson. "*The Thing From The Future.*" http://situationlab.org/project/the-thing-from-the-future/

10 Khemo is the brand name given to the artifact by the workshop participant that created the manuscript.

9

FUTURE HERITAGE

A Community-Based Exchange between Berlin and Ramallah المستقبل تراث

Jasmin Theresa Grimm and Sally Abu Bakr

The Notion of Heritage

Often we speak of heritage. 2018 was known as the Year of European Cultural Heritage. But what actually comes to mind when we think of heritage? We might think of officially labelled heritage, like UNESCO World Heritage Sites, places with cultural significance, such as Galapagos Islands in Ecuador, a unique living museum and showcase of evolution, or other sites with significant landscapes or biodiversity. Man-made heritage implies tangible culture such as buildings, objects or artefacts, but heritage is not limited to material objects that we can see and touch. Languages, traditions, oral histories, social practices, traditional craftsmanship, rituals and knowledge are transmitted from generation to generation within a community as part of their intangible culture. Heritage acts as a bond between people.

In dealing with heritage, however, we often tend towards protectionism, aiming to preserve the original form as it links to a certain region while providing the basis for national history and prestige. Promoting national heritage can bring economic benefits as a draw for tourism, but political factors become a consideration. Mona Lisa at the Louvre, one of the world's most visited artworks, for example, raises questions of ownership: in this case, national ownership, as DaVinci was Italian and the painting resides in France.

Cultural objects contain knowledge that must be mediated and made accessible to publics. Countries may undertake legislative measures to prevent the exodus of certain masterpieces from their territory. These objects represent a nation's history and identity while forming a bond to the past, present and future. Although states strive to protect their heritage, we know that throughout history, colonialism, political conflicts and war have made this a challenging task.

Characteristics of national cultural heritage include historic or artistic significance, a combination of the artifact and its context, inherited through generations and passed on to the future. In this way, heritage is dynamic and changes over time; it is in principle, limitless. If we consider that the present creates the heritage of the future, then heritage can be continuously accessed, shaped, modified and disrupted in an iterative process.

Cultural Heritage in Times of Transformation

The notion of heritage becomes especially important when we examine world heritage that is in danger, such as heritage sites and traditions that are fading due to migration, politics, climate change, economic decline or war-torn environments. Artists and researchers have recently begun to explore uses of electronic media and technologies as methods to restore heritage in new forms.

- Archaeologists at New York's Metropolitan Museum of Art[1] have digitally restored artworks from an ancient Egyptian temple, showing the time of the pharaohs in a detailed and colorful light. This brings to life ancient and forgotten histories of Egyptian Pharaohs by re-applying the original colors through projection mapping.
- A similar approach was pursued in a project that used 3D projection mapping to restore Buddha sculptures in Afghanistan at their original site. The giant Bamiyan Buddhas of Afghanistan have been rebuilt using 3D light projection, filling the empty cavities where the Buddhas once stood.[2]
- In 'Nefertiti Hack,' artists Nora Al-Badri and Jan Nikolai Nelles 3D-scanned the original Nefertiti at Neues Museum Berlin[3] (Nefertiti was stolen from Egypt 100 years ago and is currently exhibited in Berlin, preventing the statue from returning to its country of origin). In a radical act, the artist duo released the data of Nefertiti so that anyone around the world can access, study, print or remix a 3D dataset of Nefertiti's head in high resolution. This gives ownership and accessibility of the object to anyone. The Neues Museum itself does not allow any access to the head of Nefertiti nor to the data. 'With the data leak as a part of this counter narrative we want to activate the artefact, to inspire a critical re-assessment of today's conditions and to overcome the colonial notion of possession in Germany'[4] (Al Badri and Nikolai Nelles).
- With the project #NEWPALMYRA[5] the artist Bassel Khartabil aims to preserve the endangered cultural heritage of Syria by creating a virtual copy. Anyone can upload images or 3D models to help restore historic sites, or as he says, fight against ISIS deconstruction.
- Morehshin Allahyari's 'Material Speculation'[6] also aims to revive ancient heritage using modern technologies. With 3D modeling and printing she has recreated artefacts that were destroyed by ISIS. In this way, she is digitally archiving and restoring history as an act of resistance.

These projects apply new methods to relevant aspects of heritage by addressing its preservation and articulation or by redistributing ownership. Practices such as these become especially crucial in the context of transformation; they target cultural artefacts, knowledge and sites that are inaccessible or fading while reviving discourse around heritage through the use of new media and technologies.

Future Heritage // Ramallah – Berlin

In a one-year collaboration, Sally Abu Bakr and Jasmin Grimm developed a methodology for exploring identity and heritage between two cultures in the digital age. 'Future Heritage' became the curatorial theme for a German-Palestinian cultural exchange in which German and Palestinian media artists met Palestinian craftswomen and craftsmen. The project links cultural traditions to global opportunities and maker culture to craftsmanship across the fields of visual arts, handcrafts and technology.

Our collaboration between Ramallah and Berlin began with a simple question: *how can we actualize traditional crafts using technology through a community-based approach?* Jasmin Grimm has a curatorial background in Media Science and Participatory Media. Sally Abu Bakr, as Director of the Cultural Department of Ramallah Municipality, is interested in highlighting Palestinian heritage and identity, especially in the area of craft. Together, we wanted to experiment with activating new media in the distribution and re-development of Palestinian heritage.

'Future Heritage' seeks to explore new ways of articulating and distributing cultural heritage within the context of the Palestinian political situation. In the Palestinian Territories, daily life is interrupted and autonomy and future perspectives barely exist. Palestinian culture is geographically torn apart. Heritage, therefore, is the element that unifies generations but is practiced daily as an antiquity, glorifying the past and neglecting the future. Memories overlay the present in order to reaffirm Palestinian existence. Slowly transforming the actual cultural heritage through craft (in our case, pottery) is an attempt to formulate and emphasize new narratives.

A Community-Based Approach

Heritage is often used by nations to mark their culture. In communities, however, the role of heritage, its narratives and ownership is more complex and integrated. Here, intangible heritage functions as an active agent in everyday life, gluing together individuals of a society in the present time. Intangible heritage – such as language, knowledge, social practices or craftsmanship – keeps its significance, not through preservation, but rather through constant reuse and application. We can view heritage in this sense not as an antiquity, but rather as a living organism that is open to access, owned by communities and reproducing itself in a cycle of constant change.

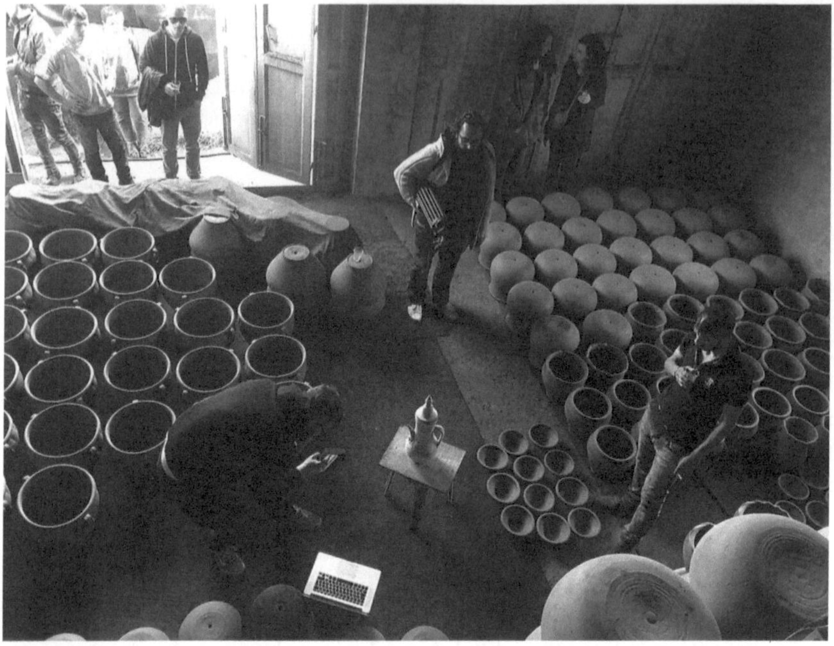

FIGURE 9.1 3D Scanning in Hebron. Image by Susa Pop.

Examining the interstice between tangible and intangible heritage and between analogue and digital, we can explore how heritage can be delivered and manifest through technology, thereby highlighting heritage dynamics in terms of distribution and actualization through community-based approaches. We ask: *In what ways can we actively and supportively design the process of cultural identity-building in a time of technological transformation?* Through 'Future Heritage,' we were interested to include new technologies for their potential to (re-)formulate heritage and thus identity. This led us to a second question: *in what ways can new technologies be embedded in the production and articulation of intangible heritage?*

Artistic Approach

The artistic process of 'Future Heritage' applies an additive method of research: local communities (potters from Hebron and Bethlehem) create new objects together with German and Palestinian artists. These objects were based on traditional forms and enhanced through digital manufacturing technologies.

At this point, a 'Call for Projects' was launched, inviting German and Palestinian artists to develop community-based artistic scenarios and methods for exploring junctions of heritage and new media.

Christian Zöllner, from the Berlin-based design collective 'The Constitute,' developed an artistic scenario for 3D printing with ceramic clay; existing craft designs would be digitized (3D scanning), transformed (3D modeling) and reproduced with digital fabrication methods (3D ceramic printing). New hybrid forms and objects would evolve that would oscillate between analogue and digital. This process of merging, multiplying and virtualizing existing cultural heritage is unique as it aims to digitally archive intangible heritage while opening space for new designs. The project also aimed to create space for international maker communities, as we intended to make the digital files and designs available in an online database for future reproduction. We thereby sought to address aspects of ownership and redistribution. We will return to this point later as we describe the project's process.

Old Traditions and New Technologies: An Experiment between Worlds

The core of 'Future Heritage' was a five-day Prototyping Lab to which we invited five local potters from Hebron and Bethlehem who practice traditional craftsmanship. Two Palestinian and one German digital artist worked collaboratively with the artisans, exchanging context, knowledge and co-creating new forms, objects and patterns.

The five-day Prototyping Lab began with field trips to Hebron and Bethlehem. It was important for us as organizers that the process start within the communities – in their work and living spaces. Visiting the artisans' workshops placed their valuable practice at the center and set the groundwork for the collaboration. While in the workshops, we could not only see their context of living but also understand more profoundly the role of handmade crafts in their families and communities.

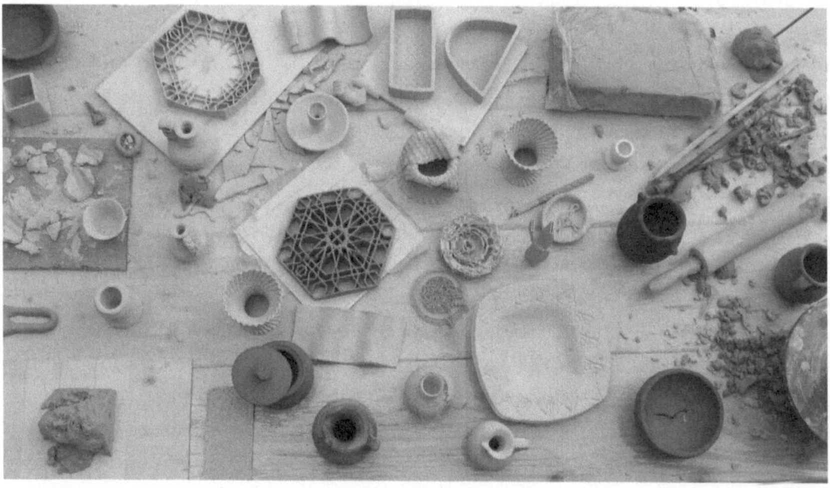

FIGURE 9.2 First 3D printed objects at Workshop in Ramallah. Image by Jasmin Grimm.

Some of the potters had inherited their craft skills over generations: from their parents, who themselves had inherited the skills and tools from *their* parents. Their way of working with clay was therefore very traditional. All production steps occurred on site: sourcing the clay from local grounds, processing it (drying, kneading, etc.), modeling it on the potter's wheel and firing the objects. Everything happened in and around the workshops. The artists used the field trips to create initial models with a handheld 3D scanner that allowed the pottery objects, such as vases, jars, etc. to be digitized in a first rough step. This also allowed the potters to slowly approach the technological aspect of the project, though there remained a bit of skepticism.

Following the field trips, all participants met in the newly opened Hosh Qandah Community Center in the old city of Ramallah. In the next days, a level exchange took place; the potters taught the media artists how to use the potter's wheel and create clay objects from scratch. Here, the difficulty of crafting center-based objects became clear, considering that the thickness of the wall, symmetry and timing all played major roles in successfully shaping clay. The potters, on the other hand, learned about digitization processes. The handmade objects of the potters, which were 3D scanned, could be modified and disrupted through 3D modeling software. After remodeling the forms, the 3D objects were printed in clay on-site with a special 3D clay printer. The objects, thereby, were returned to their material of origin and the practicality of 3D scanning and 3D modeling was revealed.

Digital Craftsmanship

Through experimental means, we collaboratively explored possible directions for actualizing intangible heritage in a highly tangible way. We aimed to connect craftsmanship to maker culture, local needs to global innovation and traditions to new technologies. Using 3D printing with clay allowed us to combine the knowledge of two worlds: maker tools and traditional pottery. The process was new to both sides. The artists don't usually work with one distinct technique or craft, acting more often as interface agents. This process helps them to adapt to new ideas. At the same time, the craftswomen and craftsmen are experts within their specialized techniques, repeating and mastering the pottery process over years of practice.

There was both skepticism and curiosity when the two groups were brought into exchange. The potters were resistant to the technology, as they feared it could literally cost them their job and replace their long-developed skills. Still, curiosity prevailed and once they saw exactly how the 3D clay printer functions, they were surprised at how delicate, time intensive and error-prone the machine was. One potter joked, saying it would be more efficient to put a potter from Hebron into the machine. This immediately increased the potters' confidence and perceived value of their handiwork, as indeed, they are much more efficient at producing clay objects with the potter's wheel than the 3D printer is at manufacturing them. Market-wise, the machine is no direct competitor, nor will it

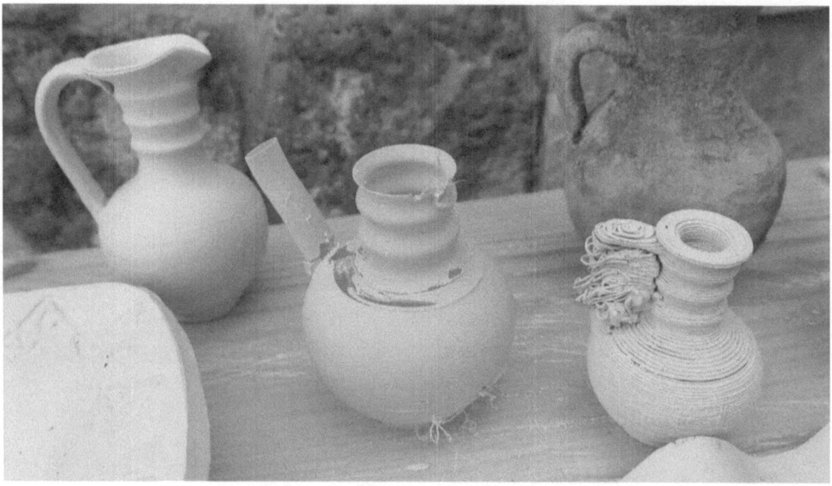

FIGURE 9.3 Hand pottered and 3D printed objects in comparison. Image by Jasmin Grimm.

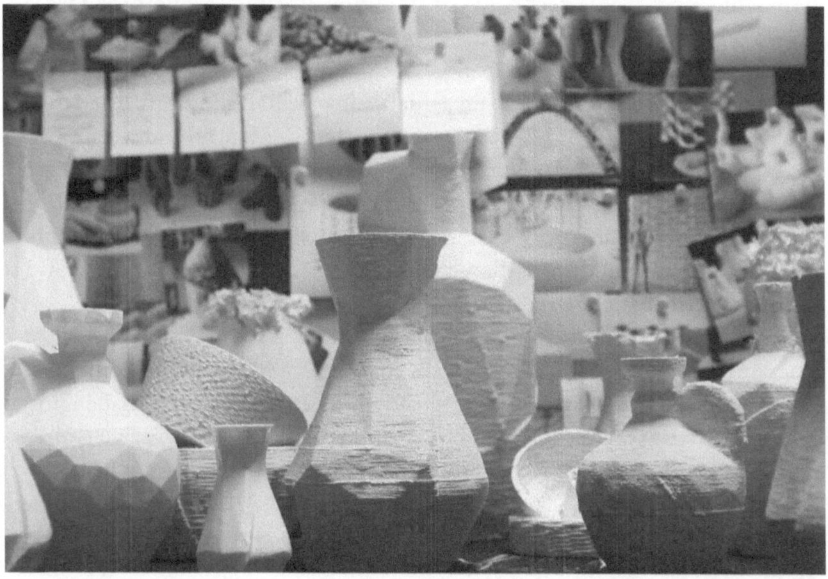

FIGURE 9.4 3D-printed and fired clay objects. Image by Christian Zöllner.

replace their craft skills. The potters, however, saw the potential of developing new designs that are simply not possible to be manufactured by hand, such as merging two objects into one. In this way, digital manufacturing processes can be seen as adding to the existing skills of craftswomen craftsmen. The digital artists, on the other hand, could learn how clay, quite a slow material, works. Using digital tools can be fast-paced and far removed from material processes. 3D ceramic printing is a delicate hybrid between old techniques and new technologies. While being comfortable with using the machine, the artists had to learn how to shape clay, a time-sensitive material that needs to be mixed to a certain consistency, dried at the correct speed and fired using traditional techniques. The 3D clay printing might be a new technology, but along all of the steps of working with clay, the potters' knowledge and intuition are crucial. This again shows the value of traditional knowledge, as it is an elementary component in digital transformation. The machine is not a robot that produces large amounts of pots in an assembly-line style. It is rather a tool for the development of prototypes, new forms and designs. In this way, we could call it a 'digital manufacturer' that combines digital skills with traditional knowledge.

New Technologies / New Territories?

Technologies contain inherent potential to transform existing processes of reproducing and delivering culture, identity and thus, heritage. By combining crafts with new media technologies, traditional forms can be produced and shaped – beyond physical borders!

'Future Heritage' aimed to explore how heritage can migrate beyond territory, connecting communities and actualizing cultural traditions. Viewing the project in the context of Palestinian Territories, these new digital fabrication methods also open up new possibilities for spreading heritage that is detached from place. By uploading the designs into the Cloud and virtualizing Palestinians' tangible heritage, it can be spread beyond physical borders. The digitally archived objects would then allow communities in other parts of the world to be part of a process of rearticulation.

Our initial idea was that all collected information, such as 3D data files and programs, would be open sourced to allow a distributed community-driven process. As implied earlier in this essay, this online database became challenging in light of the context of the political situation of Palestine. As its heritage is already fading, one of the greatest concerns is that Palestinian heritage is taken over through the occupation, claiming artefacts like pottery vases that are produced in the Palestinian territories by local potters as Israeli heritage. Thus, by uploading the open 3D files into the Cloud, the scans of the vases could be reproduced and thus be issued under another national identity. In this very special case, this would be an enormous issue that plays into one of the key problems in the conflict. We were therefore confronted with the dilemma of the open source and open ownership ethos versus protection and control of who actually accesses the database. With this in mind, we had a deep motivation to question to whom identity belongs in a digital era.

Key Learnings

'Future Heritage' is part of a larger initiative called Tandem Shaml, led by MitOst e.V. who foster a mutual dialogue between the Arab World and Europe. 'Future Heritage' seeks not only to expand and actualize Palestinian heritage but also to bring attention to its arts and culture. The objects developed were exhibited in Berlin during Maker Faire and Retune Festival as well as in Marseille at Villa Méditerranée. The project extends beyond creating new opportunities for Palestinians by becoming a platform for dialogue, mutual exchange and creating alternative narratives that challenge media stereotypes.

We aimed to promote a sustainable transformation of heritage with local culture at the heart, exploring artistic potential for local businesses in networked societies and building bridges between global innovation and local needs. Along the project's process, we were confronted with challenges regarding the political context that reshaped the initial concept as we learned more about Palestinian identity. The platform created between the potters and digital artists was both challenging and experimental; it was a collaboration not only between cultures but also between professions. Still, it allowed us to explore common ground while fostering future potential for participants. The digital artists from Berlin, for example, began working long-term with 3D ceramic printing and brought their new expertise to traditional clay manufacturers in Germany.

One of the biggest outcomes came from our own collaboration, that between Sally Abu Bakr and Jasmin Grimm. None of this would have been possible if we, as a cross-cultural team, would not have trusted one another. Key to this collaboration was how we tackled challenges that arose along the way: mutually and on equal footing. This was not only the greatest value within our collaboration but also how we structured the whole process of exchange: bridging cultures, communities and professions. Particularly viewing the project and process within a greater geo-political context, it is important to mention that Euro-centric or Western approaches alone will not create impact or change. If we want to foster meaningful exchange, we need to build mutual and open platforms for dialogue to learn from one another.

Notes

1 https://www.metmuseum.org/blogs/digital-underground/2015/color-the-temple
2 http://www.lionsroar.com/afghanistans-giant-buddhas-rise-again-with-3d-light-projection/
3 http://nefertitihack.alloversky.com/
4 Tyldesley, Joyce. 2018. *Nefertiti's Face: The Creation of an Icon*. Profile.
5 http://www.newpalmyra.org/
6 http://www.morehshin.com/material-speculation-isis/

10

LITTLE INVENTORS

From Artistic Method to Global Brand

Suzy O'Hara

Background

Little Inventors inspires and encourages children to build creative confidence and develop a lifelong passion for creativity, problem-solving and imaginative play through invention. We fuse the limitless imagination of children with professional 'making' practices drawn from the arts, industry and the DIY Maker Movement, showcasing children's ideas in physical and online exhibitions globally. In this chapter, I will analyse the context, drivers and trajectory of two distinct Little Inventors projects, *INVENTORS*! and *Little Miss Inventor Challenge,* via the lens of the inventions that have been imagined by our little inventors and turned into real objects by our Magnificent Makers (MM). In so doing, I hope to offer some insights into how and why we do what we do and what we have learnt so far.

I first met artist, designer and inventor Dominic Wilcox when I commissioned him to create a new work for Thinking Digital Arts, a digital arts festival that I founded in Newcastle, UK (2014). The festival allowed me to explore commissioning strategies that developed cross-sector, interdisciplinary practices and new relationships between artists and commercial creative technologists.[1] My interest in Dominic's work was initially sparked when I came across his 'No Place Like Home' GPS shoes,[2] which were inspired by Dorothy's red shoes from The Wizard of Oz. Working with a shoe maker and a technologist/maker, Dominic created a pair of shoes that can help you find your way home by linking wirelessly to networked software and GPS technology. The shoes are activated by clicking your heels together. One shoe features a progress bar to tell you how close you are to your destination, while the other shoe provides you with your direction of travel.

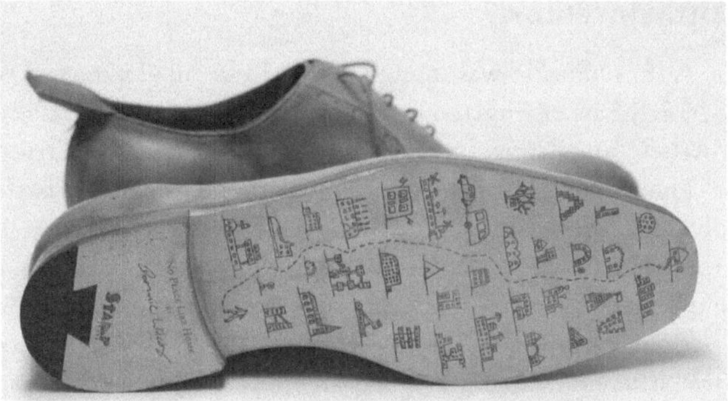

FIGURE 10.1 Dominic Wilcox, *No Place Like Home GPS Shoes*. Photo by Joe McGorty.

While many artists generally avoid working within the world of commerce, over the past two decades, Dominic seemed to have developed a healthy interest in both creative and financial opportunities that working with commercial commissioners afforded. He had particularly achieved success with commercial brands, such as BMW, Nike, Kellogg's and Paul Smith. However, while he worked across very different contexts, Dominic's practice clearly resided within values and methods that are primary to art and design. He had a keen interest in invention, which he used to catalyse his ability to imagine alternative, often humorous and always speculative solutions to perceived, everyday problems:

> Dominic Wilcox's drawings aren't just witty and beautifully drawn, they are serious challenges to the real world to keep looking at itself with innocent eyes, wondering what else is possible.
>
> *Designer, Thomas Heatherwick*[3]

His work represents a new kind of artistic practice that helps articulate the possibilities of the role and function for art within other industries and sectors. He is representative of a growing cohort of artists (others include Yuri Szuki,[4] Di Mainstone,[5] Memo Atkin[6] to name but a few) that devise creative strategies to support the development of their own expertise via commercially focused consultancy or client-based briefs. Often, this commercially focused work resources the development of their artistic practice, which ranges in the types of outputs and venues in which it is shown, from museums and galleries to festivals and events.

This balance of interests and hybrid creative practices have brought Dominic's work to the attention of an increasingly global audience. By time I met him in 2014, he was already an established artist with a global reputation in art, design and invention. He was also a sought-after inspirational speaker on creativity, innovation and how to find great ideas, giving talks to a range of audiences from diverse contexts including universities, conferences and commerce.[7]

What Is Little Inventors?

In Autumn 2015, Dominic was invited to return to his hometown of Sunderland in the North East of England by The Cultural Spring, an arts organisation funded by Arts Council England through an initiative called Creative People and Places.[8] The aim of the Cultural Spring was to 'focus on parts of the country where involvement in the arts is significantly below the national average' and 'to transform the opportunities open to people in those places.'[9] He was invited to conceive and deliver a participatory art project that would engage with young people in particular areas of the city. As such, the INVENTORS! Project[10] (the precursor to Little Inventors) was primarily a participatory arts commission, designed as a way to engage young people living in a socially and economically depressed city in the North East of England, in arts and cultural-based activity.

As an art commission, the INVENTORS! project allowed Dominic to practically explore the theory that creativity is inherent within everyone and can be usefully applied to every aspect our lives, every day. The freedom afforded by this context also enabled him to explore alternate ways of envisioning and enacting his role as 'artist' within his creative process. Contrary to his usual practice, he decided he would not generate or imagine ideas himself. Rather, he would ask children in his home town of Sunderland, to imagine new ways to tackle perceived problems or dream up ways of making particular tasks easier, more fun or interesting. We then asked local makers and manufacturers to select the strongest ideas and turn them into real things for exhibition.

Three years on, Little Inventors[11] continues to fuse the limitless imagination of children with professional and DIY 'making' practices. Our mission to nurture creativity through invention and our methods of engaging both children and adult makers has incited a global response from a diverse range of cultural, educational, business and civic partners. As a result, we have delivered invention challenges to children in over thirty countries, including Canada, China and Poland. We make and exhibit children's often extraordinary, thought-provoking and ingenious ideas in physical exhibitions across the world. At the time of writing, our online platform presents over 7,000 uploaded invention ideas, a third of which have had individual feedback from a member of the extended Little Inventors team of moderators. Our website delivers a growing suite of digital content (images, video, animated gifs etc.) to an engaged, global community.

As a creative start-up, Little Inventors has quickly expanded to China reaching thousands of children through workshops and thousands more through large-scale physical exhibitions. In 2016, the Natural Science and Engineering Research Council (NSERC) of Canada piloted LI across five cities. In 2017–2018, we delivered the Inventions for Space Challenge in partnership with NSERC and the Canadian Space Agency. We have now re-launched the Space Challenge for another year to coincide with the voyage of astronaut David Saint-Jacques to the International Space Station, inviting children to think about life in space. The strongest ideas will be chosen by our partners, Dominic and the Little Inventors team and made into real objects. The final fabricated objects will be shown on the International Space Station in 2019. We have also launched Little Inventors Middle East and are the lead children's project for the forthcoming Sharjah Children's Biennale.

Rationale for Reflecting on LI through Practice

Through our work, Little Inventors has been steadily growing a network of artists, designers, craftspeople, technologists, DIY and professional makers, scientists, manufacturers and heritage craftspeople, with whom we work to realise the inventions envisioned by the children we engage. We now call this cohort of collaborators Magnificent Makers (MMs), and we currently have nearly 120 MMs presenting a profile on our website.[12] Makers are located in Canada, China, Hong Kong, UK, Ukraine, United Arab Emirates, Switzerland, Netherlands, India and Nigeria. They work within different geographical and social contexts, across many different sectors and industries and often engage with interdisciplinary and cross disciplinary strategies within their practices. The following objects, imagined by our little inventors and made by our MMs, offer some insights into our programme and what we have learnt so far.

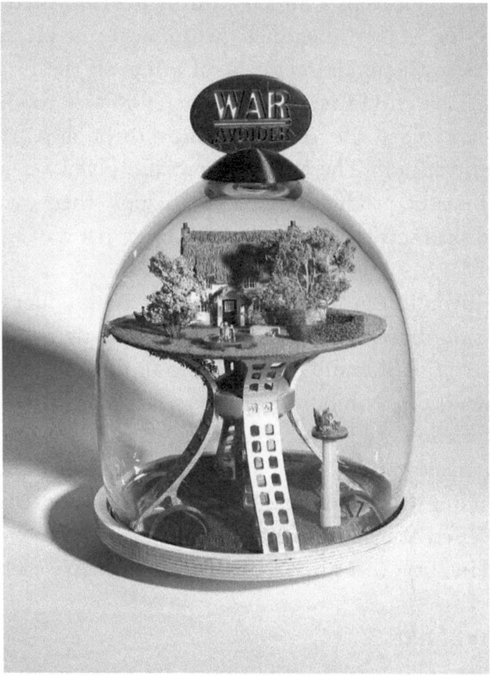

FIGURE 10.2 *War Avoider.* Photo by Colin Davidson, Rosella Studios.

Case study 1: Artist Commission

Project: *INVENTORS!*, 2015–2016
Invention idea: The Liftolater (War Avoider)
Inventor: Charlotte, age 11
The Maker: Erin Dickson

The Project: INVENTORS! with Dominic Wilcox

Finding and sharing good ideas was the key motivation behind the initial INVENTORS! project. In order to access and engage as many children as possible, we facilitated 19 invention-focused workshops, in schools and community venues across Sunderland. Over the course of two weeks, we gathered just over 600 invention drawings from children aged between 4 and 12.

Next, we engaged with a diverse community of professional makers through an open call to participate. Over 60 invention ideas were shortlisted and presented to makers at FabLab Sunderland. Each maker then selected the invention object they would like to make, many of which had no precedent for production.

Rather than Dominic assuming his usual creative role and leading the design and production process, the child was allowed to supply the specifications to their Maker directly, through a facilitated series of face-to-face 'Meet the Maker' meetings. In doing so, he placed the responsibility for the aesthetic and conceptual intent of the final works with the children. Each child inventor was therefore able to claim and learn about his or her role as the 'Artist Inventor' throughout the production of their own invention. The Makers were also given a degree of autonomy throughout the production process to apply their own professional creativity and skill sets to realise the prototype in a way agreed with the little inventor.

In all, 23 invention prototypes, including 3D product drawings, animated invention drawings and all of the 600 2D drawings were shown in the INVENTORS! exhibition, which ran for two weeks in January 2016. The two-week exhibition or 'ideas lab' was hosted in an empty shop in Sunderland City Centre and attracted over 1,000 visitors.

The project achieved worldwide attention with national and international coverage including CNN International, The Discovery Channel, CBBC, The Times, Wired Magazine and Its Nice That, an online, creativity platform founded in 2007, that reaches over a million people each month.[13]

The Invention Idea

Charlotte, aged 11, imagined her invention idea in an invention workshop in her primary school in Sunderland. On her drawing, she describes her invention idea as a way to 'get away from war' by simply raising your house and garden above it, using a 'liftolater.' Protected by an invisible blanket, you then steer your house to a safer location, away from the conflict zone.

Charlotte's idea was collected in paper form and chosen as one of 60 shortlisted ideas that would potentially be made. All shortlisted ideas were then scanned and collated into a PDF document for sharing with makers. Artist and fabrication technician Erin Dickson chose to make her idea and met Charlotte at a 'Meet the Maker' event to discuss the design details of her idea before it was realised.

The Maker: Erin Dickson

Erin Dickson is an artist and digital fabrication specialist and helped materialise Charlotte's creation at FabLab Sunderland. Fab labs are hosted in a variety of commercial, educational and cultural contexts and are becoming hubs for activities that bring academic research, creative practices, commercial innovation and a STEM-focused schools education programme together in the one location.

As an artist, Erin's practice combines 'architecture and digital technology to consider the emotional qualities of urban and domestic spaces.'[14] Her particular fusion of artistic practice and research and her digital fabrication expertise had a significant impact on the quality and rendering of the final aesthetic and ultimate reading of the War Avoider. Her creative interpretation of the War Avoider[15] was as an object to observe, rather than a functioning prototype to interact with. However, the quality and breadth of the making skills and the critical consideration of aesthetic helps to articulate Charlotte's invention idea, using a fusion of interdisciplinary skills and influences. War Avoider was one of five INVENTORS! objects accessioned by the Victoria and Albert Museum in 2016.[16]

The Learning

Alongside academic and commercial interest, the INVENTORS! call for Makers also attracted the attention of the DIY Maker, artistic and craft-based communities. A broad breadth of artists and craftspeople, including theatre designers (Theatre Unlimited), product designers (Solid Ideas), digital artists (Lalya Gaye, Attaya Projects[17]), textile artists (Kate Eccles) as well as traditional glass makers (Wearside Glass) and DIY makers (MakerSpace Newcastle[18]) also applied their particular skill sets to creating invention objects for the project.

INVENTORS! provided an opportunity for these more creative disciplines to develop their own practices, try new ways of working in a creative context and generate new ideas themselves. While each of the

(Continued)

Makers was asked to stay as true to the original drawing as possible, each of the different communities of practice brought their own particular production style and design aesthetic to the final invention prototype that they made. In this respect, each invention object was bespoke to the particular pairing between the child inventor's idea and specifications and the Maker's background, values, process for production and their personal motivations for choosing the invention idea. The resulting exhibition reflected this range of values and methods and was curated to feel more like an ideas laboratory than a fine art exhibit.

The global publicity that surrounded the first INVENTORS! alongside Dominic's growing reputation as a notable contemporary designer, brought the project to the attention of Corinna Gardner, Senior Curator of Design and Digital at the Victoria and Albert Museum (V&A). Corinna went on to accession five invention objects (imagined by children in Sunderland and made by local North East makers), into the permanent collection of the V&A. The acquisition of the five INVENTORS! objects highlights the perceived cultural value of the project, the ideas and objects. It also underlines Dominic's art and design practice from within which the project was conceived.

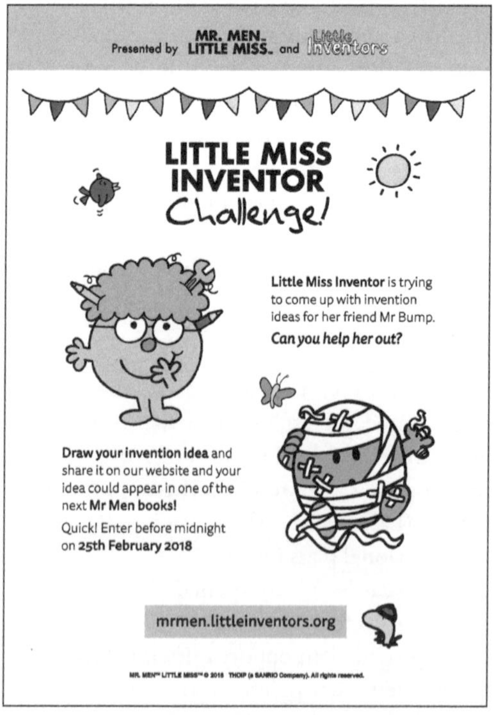

FIGURE 10.3 *Little Miss Inventor Challenge!* © MR MEN™ LITTLE MISS™.

Case Study 2: Engaging with Brands

Project: Little Miss Inventor Challenge, 2018
The invention idea: Air Bandages
The inventor: Beatrice, aged 5
The Maker (Illustrator): Mr Men Little Miss Artists

The Project

In late 2017, Little Inventors was contacted by Egmont, the largest specialist children's book publisher in the UK. Egmont was launching the 39th Mr Men Little Miss character,[19] Little Miss Inventor, on 8 March 2018 (International Women's Day), with Sanrio,[20] the Mr Men Little Miss brand management agency. They invited Little Inventors to support the launch of their new character by delivering a competition aimed at four- to six-year-old children across the UK. The strength of synergies between our brands and the global profile and reach of Mr Men Little Miss books within our target market for engagement, made it an exciting proposition.

The competition or challenge launched at the end of January 2018 and was supported by a national PR campaign, delivered by Aduro Communications. We invited children in the UK to think up invention ideas to help Mr Bump. We were given three weeks to reach a target of 200 invention 'entries' from our target age range. Children were directed to the dedicated Little Inventors microsite, mrmen.littleinventors.org, to find out about the competition, download a challenge resource pack and upload their idea to enter the competition. The prize for the winning invention idea was to have their invention illustrated by a Mr Men Little Miss Artist, (illustrators employed to work in the style of Adam Hargreaves) and appear in an official Little Miss Inventor storybook in the upcoming series. The winner of the challenge and five runners up received illustrated drawings of their invention ideas at a Little Miss Inventor launch event at Manchester Museum of Science and Industry (MSI).

Dominic attended the launch event at MSI and presented the winners with their prizes. The partnership between Mr Men Little Miss and Little Inventors garnered some national coverage, including the *Yorkshire Evening Post*,[21] *The Telegraph*[22] and the *Manchester Evening News*.[23]

The Idea

Over 800 children entered the Little Miss Inventor competition over the course of three weeks. The winning entry was by Beatrice, aged five years, who invented Air Bandages that would keep Mr Bump safe. The

(*Continued*)

bandages were designed to fill with air and expand so they protected Mr Bump when he bumped into something. Beatrice's entry was drawn, then uploaded onto the dual branded microsite by her parent or carer. Her now digital invention drawing received individual (positive) feedback by one of the LI team of moderators. As a way to facilitate positive public engagement on the website, all invention drawings uploaded onto the Little Inventors website are open to receiving positive feedback from website visitors, who either 'like,' 'love' or find inventions 'funny.' In all, Beatrice's invention garnered 441 positive responses.

Her invention idea was shortlisted by Dominic and our Chief Educator Katherine Mengardon, before being selected as the winning entry by the team at Egmont.

The Maker (Illustrator): Mr Men Little Miss Artist

The prize for the winning invention idea was to have their invention illustrated by a Mr Men Little Miss Artist and appear in an official Little Miss Inventor storybook. In all, four invention drawings were illustrated in the Little Miss style, including Mr Bump's Brilliant Bandage Put-A-Onner, the Flinger Winger Suit, the Crazy Cutter and Mr Bumps Marshmallow Machine.[24]

As the invention was drawn in the Mr Men Little Miss brand style, by a 'house style' artist, the name of the Mr Men Little Miss Artist who illustrated each invention drawing was not revealed. Unlike INVENTORS!, it was the Mr Men Little Miss brand aesthetic, values and methods rather the individual artist practice or identity that took precedence over the final outcome.

Learning

There are different kinds of value that working with brands can offer, financial, reputational and learning. Interestingly, there was no budget to pay Little Inventors for delivery of this particular brand-led competition. Therefore, much like artists who are often asked to deliver work for little or no budget, the team had to decide if the delivering the Little Miss Inventor Challenge at a loss was worth our investment.

The project provided a significant reputational opportunity for us to align ourselves with a brand of children's books known throughout the world. Being a relatively young brand, the opportunity to align ourselves with such a well-known name that had great synergy with our mission was a primary factor in our decision to say yes. However, other significant factors included the opportunity to further develop our offer and focus upon a four- to six-year age group, as it allowed us to engage with a much younger inventor than our usual programme; the opportunity to develop a literacy-based resource suitable for this age group. This resource could (and would) be used as a template for other curriculum-based resources we were developing for other (paid) projects. The output was a book rather than an exhibition, offering us a format that was easily distributable and accessible to our audiences.

The project also showed how artists can engage with brands in a transactional, rather than collaborative way. The relationship between Mr Men Little Miss artists and Egmont was a stark contrast to our relationship with Andy Mattocks, a 3D product designer who had illustrated a suite of product drawings for the INVENTORS! Project final exhibition. Whereas the Mr Men Little Miss artists were contracted to use the brand's aesthetic and style, Andy had been given full creative freedom to visualise the ideas in his own style, using his own methods.

Discussion

Little Inventors raises some interesting questions about the historically contentious relationship between art and commerce from within a 21st century context that boasts a thriving creative economy, an exploding maker culture and a global thirst for innovation.

The INVENTORS! Project and the Little Miss Inventor Challenge are just two case studies within a growing portfolio of workshops, challenges and projects delivered within museums (V&A, Museum of London), hospitals (Great Ormond Street Hospital), heritage venues (Alexandra Palace (Ally Pally)), government funded festivals, such as Great Exhibition of the North[25] and online, in partnership with brands (Ocado and Aquafresh). Our eclectic, ambitious and often times reactive programme of activity has proven to be an ambitious provocation that makes a clear case for the value of artistic strategies within commercial and non-cultural contexts, while opening up these contexts as potential new sites for creative practice to develop.

Artists have long been recognised as possessing a tacit understanding of the productive synergies between critical observation, reflective creative practice and generating original knowledge and innovative ideas.[26] As the search for innovative and alternative methods of working has intensified, creativity and the skills of artists' hold new value.[27] Little Inventors has challenged perceived distinctions between sites of artistic and economic production, by brokering new, direct ways of working between these often disparate sectors and industries. Deeply ingrained bias of judgement, informed by traditional understandings of the role and place of the artist in society[28] are laid bare as artists, like Dominic, and artists like him, boldly assert their agency within commercial (not for profit) business models, such as Little Inventors.

However, this is not without its challenges and in recent years, there has been some good work that explores how the arts are now more willing to engage with economic perspectives much more readily than before.[29] There has also been a breadth of work analysing the connections between the subsidised, voluntary and commercial culture and the arts role in generating value through creativity that could lead to innovation and economic benefit.[30]

In order to survive financially, we have capitalised upon the synergies our projects have with these particular fields of thought. In order to work with private and commercial partners, we have had to; crystallise our aims, refine our delivery, build our brand, adapt our offer to a spectrum of cultural, educational, commercial and civic agendas, timelines and budgets via an operational lens that foregrounds efficiency and scale. Naturally, as operational pressure to deliver and financial pressure to survive has increased, it has been a growing challenge to embed opportunities for the team to engage in more time intensive, artistically – led activities (such as critical observation, creative experimentation and reflective practice). Unsurprisingly, opportunities for our creative maturation as a company have primarily been facilitated by public arts and cultural funding and publicly funded, commission-based projects that value artistic development over more commercial concerns. Our resilience and capacity to operate and grow in our current ecology requires a delicate balance between the two to be maintained.

Conclusion

Little Inventors has proven to be an ambitious provocation that makes a clear case for arts-led intervention within STEM-focused educational agendas and innovation and commercial modes of production. While these contexts are often driven by motivations different (and often times contrary) to those of the arts, this chapter points to an evolving fluency and confidence within artistic practices that are successfully operating across cultural and commercial disciplines, sectors and industries. Little Inventors articulates how the core values and working methods of art, design and making practice can both benefit from and inform alternative contexts for creative production in tangible ways.

Notes

1 O'Hara, Suzy. 2016. "Collaborations between Arts and Commercial Digital Industry Sectors: A Curatorial Practice-Led Investigation of Modes of Production." Doctoral Thesis, Sunderland: University of Sunderland.

2 Wilcox, Dominic. No Place Like Home, GPS shoes. http://dominicwilcox.com/portfolio/gpsshoe [Accessed 20 Nov. 2018].

3 Heatherwick, Thomas in Biography of Dominic Wilcox. http://dominicwilcox.com/about/ [Accessed 20 Nov. 2018].
4 Szuki, Yuri. 2019. Yuri Suzuki | Home. Yurisuzuki.com. http://yurisuzuki.com/ [Accessed 20 Nov. 2018].
5 Mainstone, Di. 2019. Ambassador for Sennheiser's Momentum Campaign. http://dimainstone.com/project/sennheiser-momentum/ [Accessed 12 Jan. 2019].
6 Atkin, Memo. Works. http://www.memo.tv/ [Accessed 12 Jan. 2019].
7 Wilcox, Dominic. Talks/workshops. http://dominicwilcox.com/talksworkshops/ [Accessed 12 Jan. 2019].
8 Creativepeopleplaces.org.uk. Creative People and Places. http://www.creativepeopleplaces.org.uk/ [Accessed 12 Jan. 2019]. CPP is a national program, that focuses investment in parts of the country where people's involvement in the arts is significantly below the national average. In 2015, the city was ranked in the top 20 UK local authority districts with the highest levels of income and employment deprivation.
9 Artscouncil.org.uk. Creative People and Places: Round 1 extension | Arts Council England. https://www.artscouncil.org.uk/funding/creative-people-and-places-fund#section-2 [Accessed 12 Jan. 2019].
10 Wilcox, Dominic. 2016. INVENTORS! Project. http://inventorsproject.co.uk/ [Accessed 12 Jan. 2019].
11 Ibid.
12 Littleinventors.org. 2019. Magnificent Makers. https://www.littleinventors.org/makers/ [Accessed 20 Nov. 2018].
13 Pritchard, Owen, 2016. Beyond the Fridge Door: Dominic Wilcox Unleashes the Design Capabilities of Kids. *It's Nice That.* https://www.itsnicethat.com/articles/dominic-wilcox-inventors-workshop-280115 [Accessed 20 Nov. 2018].
14 Dickson, Erin. Erin Dickson: About / Work. https://www.erindickson.co.uk/about [Accessed 27 Nov. 2017].
15 Littleinventors.org. The Liftolator (war avoider). https://www.littleinventors.org/ideas/the-liftolator-war-avoider/details [Accessed 12 Jan. 2019].
16 Littleinventors.org. Victoria & Albert Museum Collects Five Little Inventors Inventions! https://www.littleinventors.org/press/victoria-albert-museum-to-own-five-little-inventors-inventions [Accessed 26 Nov. 2018].
17 Applied Digital Creativity, http://www.attayaprojects.com/ [Accessed 12 Jan. 2019].
18 The north-east space for creative geekery. http://www.makerspace.org.uk/ [Accessed 12 Jan. 2019].
19 Mr. Men. https://en.wikipedia.org/wiki/Mr._Men [Accessed Dec. 9, 2018].
20 Sanrio. https://www.sanrio.com/pages/other-countries [Accessed 27 Nov. 2018]
21 Andrew, Laura. 2018. Little Miss Inventor – An Introduction of a Strong Female Character in the New Mr Men book. *Yorkshire Evening Post*, February 26, 2018. https://www.yorkshireeveningpost.co.uk/news/little-miss-inventor-an-introduction-of-a-strong-female-character-in-the-new-mr-men-book-1-9036385. [Accessed 7 Dec. 2018].
22 The Telegraph. 2018. Mr Bump Gets Pumped: Children's Competition Entry Gives Accident-Prone Character Inflatable Bandages. https://www.telegraph.co.uk/news/2018/03/16/mr-bump-gets-pumpedchildrens-competition-entry-gives-accident/ [Accessed 7 December 2018].
23 Gill, Emma, 2018. Mr Men's New Little Miss Inventor Character Is Being Launched at FREE Museum of Science and Industry Event. February 26, 2018. https://www.manchestereveningnews.co.uk/whats-on/family-kids-news/new-little-miss-launch-manchester-14339075. [Accessed 7 December 2018].
24 Mr Men, Little Miss & Little Inventors, 2018. https://mrmen.littleinventors.org/ideas. [Accessed 28 November 2018].

25 The Great Exhibition of the North, 2018. https://en.wikipedia.org/wiki/The_Great_Exhibition_of_the_North. [Accessed 7 December 2018].
26 Harris, Craig (ed). 1999. *Art and Innovation: The Xerox PARC Artist-in-Residence Program*. Cambridge, MA: The MIT Press.
27 AHRC.UKRI. 2015. *The Cultural Value Project*. https://ahrc.ukri.org/research/fundedthemesandprogrammes/culturalvalueproject/ [Accessed 12 Jan. 2019].
28 Schein, Edgar H. n.d. "The Role of Art and the Artist." *Organisational Aesthetics* 1 (1–4). Accessed Jan. 23, 2019. https://digitalcommons.wpi.edu/oa/vol2/iss1/1/ [Accessed 12 Jan. 2019].
29 Donovan, Claire. 2013. A Holistic Approach to Valuing Our Culture: A Report to the Department of Culture, Media and Sport. Measuring Cultural Value (Phase Two). https://www.gov.uk/government/publications/a-holistic-approach-to-valuing-our-culture [Accessed on 12 Jan. 2019].
30 Sapsed, Jonathan and Paul Nightingale. 2015. Brighton Fuse, Ecology of Culture and Growth of the Arts. https://ahrc.ukri.org/documents/project-reports-and-reviews/the-brighton-fuse/ [Accessed 12 Jan. 2019].

+ **CONTEXT IV**

Hack Events, Residencies and Workshops

11

CYBORG ARTS CO-LAB

Interdisciplinary Collaboration Enriched Through Art-A-Hack™ Practices*

Ellen Pearlman

Introduction to the Cyborg Arts Co-Lab

In 2017, I was invited by Parsons School of Design to run a semester-long co-lab using methods I developed through previously produced Art-A-Hack™ events. Thus emerged the world's first co-lab in Cyborg Arts. Like Art-A-Hack, this lab facilitated creative collaborations while privileging art, creative speculation and process over the hack, or technical exactitude, leading participants to create something new. Art-A-Hack has been produced for over five years through a series of open calls and sponsorships. Its structure is influenced by Bruno Latour's Actor Network Theory (ANT),[1] which describes how to facilitate and manage creative collaborations.

Parsons' existing co-lab offered a forum in which Art-A-Hack methodologies could be applied to an extended co-creation event involving students alongside arts, media and technology professionals across a sustained 16-week collaboration. The co-lab produced incremental breakthroughs by starting, stopping, abandoning and then resuming the building of proof-of-concept artworks. This type of cycle is typical of the process employed by Art-A-Hack teams.

* Elements of this chapter were first published in the article "A Co-Lab on Developing Cyborg Arts – Interdisciplinary Collaboration and Practice Based Solutions" in VOL. 6, NO. 1, Journal of Problem Based Learning in Higher Education, 2018.

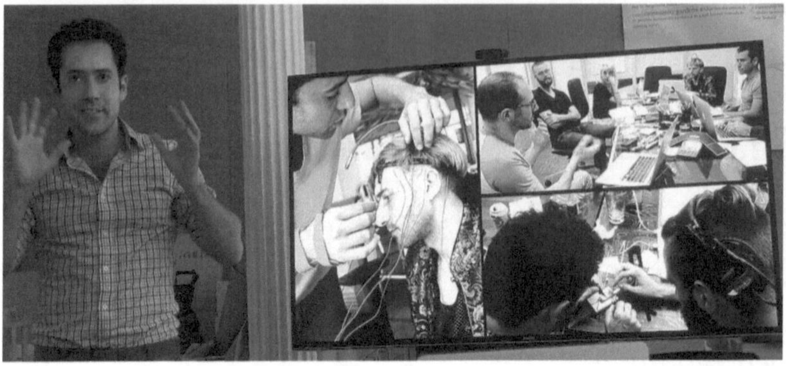

FIGURE 11.1 Engineer Oryan Inbar lectures to co-lab on creating 'heat sense' for cyborg Neil Harbisson during ThoughtWorks Arts Residency. Photo by Ellen Pearlman.

Structure and Implementation

Structuring the multi-layered co-lab required a considered curatorial process. The Cyborg Foundation, which is run by actual living cyborgs Neil Harbisson and Moon Ribas, was brought on-board for their expertise in the co-lab's central theme. Posthuman scholar Katherine Hayles defines the cyborg as someone who has 'informational pathways connecting the organic body to its prosthetic extensions.'[2] Harbisson, born color blind, implanted a sensor into his skull that turns colors (including infra-red and ultra violet) he could not see into sound, a process he uses in his creative outputs. Ribas has chips implanted in her arm and feet that sense earthquake data 24 hours a day, which she uses to create live time choreography.[3]

The goal of the Cyborg Arts co-lab was to produce a tangible proof-of-concept artwork that could theoretically be turned into a cyborg sense. The results would be envisioned to be implanted inside the living tissue of a human animal but would only be built as a proof of concept. To locate the best participants for the lab, an open call was placed in targeted venues and technology user groups in the New York area proposing a 'collaboration between artists, technologists, designers, engineers, makers and/or scientists to create and develop technologies that expand human capabilities and perception.'[4] This type of open call is typical of the way in which teams are aggregated in Art-A-Hack, although here the call was seeking individuals interested in the theme of enhancing senses using digitized parts.

More than 70 people responded to the open call, and the top project ideas were selected to create three artist-led teams; Team Glass, Team Radiation and Team Haptics. Team Glass, headed by glass artist A. focused on a sense to detect changes in the sun's solar flares. Team Radiation, led by artist D. aimed to develop a sense to detect organic and inorganic radiation.[5] Team Haptics, headed by artist S. decided to fashion a cyborg sense to coordinate her gait, which had been damaged through surgery and radiation to treat a tumor on her spine. After these teams and themes were in place, the Parsons students who enrolled in the co-lab joined the team that most appealed to them. Finally, 25 additional professionals from the open call, including programmers, research scientists, designers, artists, indie makers and non-profit executives, also joined the teams. Students were treated as fully engaged apprentices in a transdisciplinary problem-solving environment. They logged their research on a private Parsons-related Tumblr, and also posted their team's progress on the co-lab's group Slack. These digital platforms allowed me as facilitator to monitor group dynamics, momentum and setbacks in terms of ANT analysis.

Setting: Co-Lab Vs. Art-A-Hack

Parsons' collaborative laboratories allow external professionals and invited guests to participate in a normally closed academic setting. A co-lab differs from a workshop, which is usually a skill-based short-term exploration of a particular set of tasks. While students can join Art-A-Hacks outside of school time, the scenario does not depend upon an academic partner but rather various art and technology sponsors. Art-A-Hacks are usually four full days in length spread out over two weeks or a month. The co-lab encompassed one weekly three-hour session for 16 weeks. This format created a slower pace once university spring break and holidays were factored in.

Unlike a traditional Art-A-Hack, the students were assigned texts on the posthuman and cyborgs, as well as a paper about creative collaborations. Some students jumped right in and began contributing their coding or design skills towards the creation of cyborg senses. Other students chose to study theoretical aspects by researching and composing papers on topics related to cyborgs and artificial intelligence. The hands-on aspect of building and making was a new experience for some students, who were from departments such as Communications; however, all were encouraged to participate, drawing upon their abilities and chronicling their group's progress.

Various guests engaged with the lab via Skype or in person. This was more in line with a traditional university course, though Art-A-Hacks can have special guest visitors, which are usually Art-A-Hack alumni. Haribisson and Ribas Skyped in from Barcelona to initiate the first class and later to view and comment on the teams' final projects. Other guests included scientific researchers, other cyborg artists, cyborg start-up companies, body hacking conference organizers and directors of maker spaces, all of whom lectured on their areas of expertise and provided feedback. The teams would present their project ideas to the guests with both sides exchanging viewpoints. Because the topic of living cyborgs was new to almost all of the participants, the guest mentors supplied needed real-world examples and inspiration.

Methodologies: Actor Network Theory (ANT)

I used ANT to navigate the co-lab, as I have with Art-A-Hack. ANT portrays both human and non-human elements as equal actors and understands the failure and crisis of multiple actors in a network. It does this by employing a 'sociology of translation' with each 'actor' representing a vital link in the types of interchanges that occur between objects and individual subjects. A signal that is not processing information correctly, or computer code that is compiling with multiple errors is just as important as the communication between the two people who are trying to rectify an error. All components are actors in the network.

Latour believes that it is better to trace connections or 'associations' between controversies than to explain the controversies themselves. ANT examines the problems being tackled, the actors involved, how to make other actors interested in the situation, have actors agree with their assigned roles and make sure the delegated actors represent the situation correctly. If the actors are not in agreement, then the network under consideration ceases to function. This type of breakdown occurred in the duration of the co-lab. Latour states,

> You have to follow the actors themselves, that is try to catch up with their often wild innovations in order to learn from them what the collective existence has become in their hands, which methods they have elaborated to make it fit together, and which accounts could best define the new associations that they have been forced to establish.[6]

He notes that information technologies are equipped in such a technically sophisticated way that they allow us to trace the associations that were previously impossible to track.

Latour's use of the word 'actor' is extremely complex and loquacious. He says, "An 'actor' in the hyphenated expression actor-network is not the source of an action but the moving target of a vast array of entities swarming toward it...Action is borrowed, distributed, suggested, influenced, dominated, betrayed, translated."[7] I understand the use of the word 'actor' as any person or thing involved in an exchange, or chain of events that relates to a situation in the past, present or future that affects the outcome of that situation.

Within this context, any actor serves as an amalgamation of all the parts in a specific situation that communicate with one another. This is referred to as punctualization. Punctualization can be thought of as 'encapsulation,' a process of enclosing bits of programming code in 'capsules' that forms the basis of object-oriented programming. If the network breaks down, then the punctualization or communication breaks down, and the capsulation is broken open. This is referred to as depunctualization.

One of the difficulties of articulation in ANT is that everything can be viewed as either an actor or as part of the network. Which label is applied at what time depends upon the perspective or framing of the environment. Latour admits that many of his concepts and methodologies are actually ethnographic in nature and derived from the 'sociology of science and technology'[8] and that the central tenets of ANT come from a 'sociology of translation.' This does not mean that Art-A-Hack employs ethnographic techniques, or that it is a social science practice. It is first and foremost a creative arts practice, which is why it never takes place over a non-stop 24- or 48-hour time span. That type of pressure-cooker situation does not account for the needs of artists to have downtime and dreamtime to create.

Group Dynamics Considered Through ANT

Team Glass was most unclear about how to actually implement their cyborg sense of interpreting solar flares. Part of the difficulty was that team leader A. considered all suggestions from all members to have equal merit. This led to the team being unable to form a consensus, as all decisions were given equal weight and none was deemed outstanding enough to act upon. Team Radiation had a different style. They had a dominating team leader D. who shut down other points of view. The other members resented his dominance and refused to contribute anything further. Participants withdrew into silence, which led to no single decision being acted upon since no one could agree. Though the two teams had divergent styles (indecision vs. dominance), their outcome was similar in that neither could progress to the next step. Team Haptics had the most effective style of decision-making with team leader S. Though she considered others' suggestions and talked through their approaches, she was able to make the final decision, albeit with everyone's consent.

The most effective way to have everyone in Team Glass come to a consensus was to sit with them during class and discuss their ideas as a group. After one particularly rough patch of listening to all of their concerns and difficulties, I analyzed the situation using ANT methodology. I saw that they had no 'actor' (information or data) in that they had nothing supplying raw information for their project's goal. I suggested they consult NASA's online solar flare database to anchor their concepts. A team member then came up with a programming solution to connect the raw data from NASA to a piece of hardware that caused a small light to turn on each time the data reached a certain numeric threshold. Though it seemed like a small breakthrough, it completed the ANT network as the prototype was now comprised of both human and non-human 'actors' with the main 'actor' being the data that linked to the code. Once the team saw actual progress, they gained confidence in agreeing upon next steps. They were now 'punctualizing' and passing 'tokens' between one another and within the existing network. After getting an LED light to turn on when solar flare data was strong, the team embarked upon finding the correct grade of silicon to make a synthetic skin and encase the newly blinking LED lights. This skin would eventually be placed on the body as an exoskeleton for a cyborg sense.

Sitting with Team Radiation required delicate intervention on a one-to-one basis. Team leader D. deliberately spoke in more technical terms than the rest of his team in order to both confuse and dominate them. He viewed it as an affront to his abilities if he were directly questioned in front of the others. Only one other team member was technically knowledgeable enough to question him, which led to a very public stalemate between the two. When this stalemate occurred, 'tokens' or messages between actors ceased, causing the network to be on the verge of depunctualizing. It was necessary to re-initiate the exchange of tokens, or information, amongst all actors in order to drive the creative process forward. I sent individual emails to the two clashing members, then spoke to one of them privately before class, suggesting he reconsider his perspective. This led to the reintroduction of tokens, or the exchange of information.

A final example of the ANT methodology applied in the Cyborg co-lab occurred within Team Haptics' project, in which the group leader used her own body as the site for experimentation. Due to a medical condition caused by surgery and radiation to remove a tumor on her spine, her gait had a noticeable delay between her intention to walk and her actual leg movements. The idea was to build a portable motion capture detection system placed on her body that would alert her through either a slight haptic pressure, or audible sound to change her gait.

An ANT analysis of the situation revealed a functional dynamic between all the participants with a constant flow of 'tokens.' The team leader's body was the main 'actor.' That body was not communicating correctly with all of its sub-actants; it was not 'punctualizing' with its various parts. I suggested color coding specific points on the main actor's body with dots of different colored lights, and then filming them to assess her actual gait across a time span of one minute. This was a strategy to re-introduce punctualization between the parts of her body. It was accomplished by using portable accelerometers that interpreted the numeric of 'X' (length), 'Y' (height) and 'Z' (depth) coordinates. The team would then mathematically create a responsive formula to read the X, Y or Z body coordinates over time. This data served as the basis for re-punctualizing the coordinates of a depunctualized actor's body.

Before the conclusion of each session, team leaders stood to report on their progress and setbacks for that week. This demonstrated to the other teams that all were experiencing similar trajectories. For example, on a particular day, Team Glass may have understood what circuits to use, but their code may not have worked; Team Radiation may have connected two different pieces of hardware together, but the output was not clear and they had no way to interpret their data; Team Haptics may have been unable to coordinate their four accelerometers, but they were all in agreement about the difficulty.

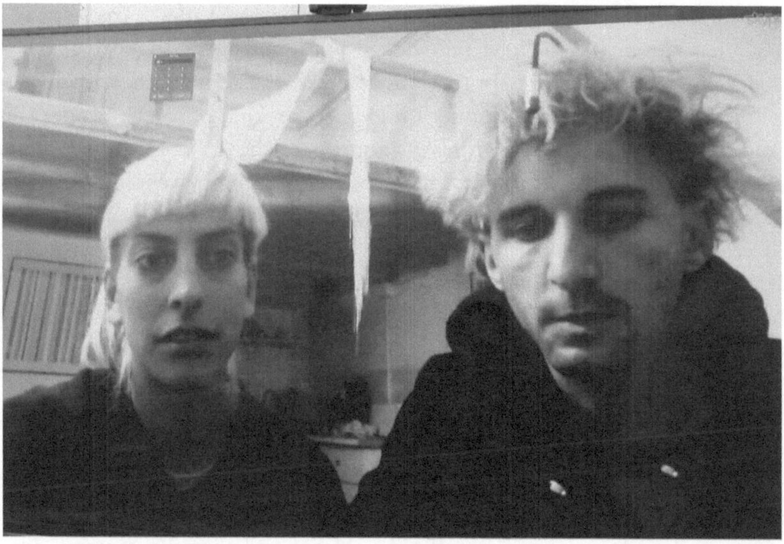

FIGURE 11.2 Cyborg Foundation founders Moon Ribas and Neil Harbisson Skype in from Barcelona to New York to speak with co-lab teams. Photo by Ellen Pearlman.

Art-A-Hack and Arts Practice as Research

How can one define arts practice as research and learning without a results-oriented investigative methodology that is quantitative or qualitative? Linda Candy, a professor of creativity and cognition research at the University of Technology, Sydney, states that this tension arises because of the need for professional practices to be defined in a way that is commonly agreed upon.[9] This commonality takes place within the confines of the research university, as opposed to other locations and institutions. The research needs to conform to those norms in order to be validated and certified as having worth, and contribute to knowledge.

Arts professor Stephen Scrivener[10] defines research as 'an original investigation undertaken in order to gain knowledge and understanding.' However, art making does not just contribute 'original knowledge' in the form of the end product art object. It is the entire process and the knowledge gained that contribute to making original artworks and spurring innovation. Scrivner argues that linguistic statements or propositions are valued in academia as contributing something of substance over art objects or creative works. The works produced by artists, such as speculative cyborg senses, do not always contain 'arguments,' the pillar of academic discourse. Because of this, arts research practice, even using methodologies like ANT, has been viewed with varying degrees of suspicion in the academy.

Curatorial and interaction design professor Lizzie Muller[11] argues that the artist/practitioner creates new knowledge while engaging in 'real situations,' instead of setting up artificial situations solely to create new knowledge. As in the Cyborg co-lab, the practitioner's role becomes that of someone adopting a 'stance towards inquiry.' New tools of inquiry are chosen from a range of practices that may involve art, design, science, engineering, psychology or critical theory to create inventive explorations within practice-based scenarios.

Creative practice does not usually begin with a hypothesis or problem, but rather, according to MIT Professor of Community Development Cesar McDowell, with an odd or 'messy' situation.[12] This is certainly the case with Art-A-Hack. A process of framing is used to locate the problem within any disorganized situation. The origins of the idea of framing arise with John Dewey's notion of the 'Problematic Situation.' McDowell explains that it begins with a 'vague image of a reality' that is identified from a surfeit of complexity.

Connecting frames through ANT analysis became a methodological solution to moments of inaction, miscommunication and system failure within the Cyborg co-lab. Action frames can be thought of as theories-in-use (op. cit.) or live responses to difficult or perplexing situations. This occurred during the building of the cyborg senses. Rhetorical frames can debate with other rhetorical frames of meaning, convincing others that a specific conceptual frame is correct. The conceptual or rhetorical frame that wins this kind of debate does so by exposing weakness in the other frame, while making sure at the same time to cloak its own weakness. Radiation team leader D. was especially skilled at this approach. Action frames occur in moments of process. They are often non-verbal and require action tasks, or motion-based changes in behavior. They may or may not incorporate the knowledge of a rhetorical-based frame, or they can derivate and create something new. This occurred most frequently with Team Haptics, which experienced the least amount of personal friction between members. Working in dynamic, evolving group situations can bring conflicts between disparate framing modalities, or can enhance these modalities. It depends upon the 'actors' within the framework.

Conclusion

Three cyborg senses as functional proofs-of-concept emerged from the Cyborg Arts co-lab at Parsons. Over the course of our weekly meetings, human and non-human 'actors' dynamically shifted within the groups. The 'actor' could be the team leader, the programming code or the hardware one week and the following week it could be any of those three designations or more than one of them. In the co-lab, the 'actor' was examined to see how it communicated (punctualized), or did not communicate (depunctualized) within the confines of the network. Included were what kind of 'tokens' were, or were not, passed. Structuring framing modes based on ANT analysis allowed groups to find different solutions. To facilitate this, I had to constantly assess group dynamics and use non-didactic interventions to keep all of the 'actors' in the ANT network fully engaged. Using this methodology in Art-A-Hack, co-labs and similar collaborative making events has resulted in dynamic solutions within a shifting matrix of professionals, students and evolving technologies.

Notes

1 Latour, Bruno. 2005. *Reassembling the Social – An Introduction to Actor Network Theory*. Vol. 7. Oxford: Oxford University Press.
2 Hayles, Katherine N. 1999. *How We Became Posthuman*. Chicago: The University of Chicago Press, 2.
3 Harbisson, Neil, Moon Ribas, and Ellen Pearlman. 2017. "Cyborg Futures | Re-Invent Yourself." http://www.cyborgfutures.com [Accessed July 29, 2017].
4 Ibid.
5 Inorganic radiation here refers to electromagnetic signals. Organic radiation, meaning a natural type of radiation, could be something as simple as a sound wave.
6 Latour, *Reassembling the Social...*, 12.
7 Ibid., 46.
8 Cressman, D. 2009. "A Brief Overview of Actor-Network Theory: Punctualization, Heterogeneous Engineering & Translation." http://summit.sfu.ca/item/13593 [Accessed May 28, 2016].
9 Candy, Linda, and Ernest A. Edmonds (eds), 2012. *Interacting: Art, Research and the Creative Practitioner*. Oxford: Libri Publishing.
10 Scrivner, Stephen, and Peter Chapman. 2004. "The Practical Implications of Applying a Theory of Practice Based Research : A Case Study." *Working Papers in Art and Design*, 1.
11 Muller, Lizzie, 2012. "Reflective Curatorial Practice." In *Interacting: Art, Research and the Creative Practitioner*. E. Candy and E. A. Edmonds (eds). Oxford: Libri Publishing, 94–106.
12 McDowell, Ceasar. 2007. "Reflective Practice: An Approach for Expanding Your Learning Frontiers." MIT OpenCourseWare. https://ocw.mit.edu/courses/urban-studies-and-planning/11-965-reflective-practice-an-approach-for-expanding-your-learning-frontiers-january-iap-2007/ [Accessed October 14, 2015].

Bibliography

Warren, Bennis, and Patricia Ward Biederman. 1998. *Organizing Genius: The Secrets of Creative Collaboration*. New York: Basic Books.

Candy, Linda. n.d. "Practice Based Research: A Guide." CCS Report: November 2006-V1.0, University of Technology, Sydney.

Candy, Linda, and Sam Ferguson (eds), 2014. *Interactive Experience in the Digital Age: Evaluating New Art Practice*. Cham: Springer Series on Cultural Computing. Springer International Publishing.

Haraway, Donna. 1990. *Simians, Cyborgs, and Women: The Reinvention of Nature*. 1st edition. New York: Routledge, 149–181.

Harbisson, Neil, 2012. *I Listen to Color*. TedX Talks, https://www.ted.com/talks/neil_harbisson_i_listen_to_color [Accessed March 23, 2016].

Haribisson, Neil, and Moon Ribas, "Cyborg Foundation." https://www.cyborgfoundation.com/ [Accessed March 23, 2014].

Hu, Jane C. "Sense Hacking: The Real-Life Cyborgs of the DIY Augmentation Scene." How We Get To Next. August 15, 2017. https://howwegettonext.com/sense-hacking-the-real-life-cyborgs-of-the-diy-augmentation-scene-e339ac9855bf [Accessed August 20, 2017].

Kozel, Susan, 2015. *Phenomenology – Practice Based Research in the Arts*. USA/Sweden: You Tube. https://www.youtube.com/watch?t=22&v=mv7Vp3NPKw4 [Accessed May 28, 2016].

Oppenheimer, Robin. n.d. "How Does Creative Collaboration Look and Feel? The Beginnings of a Definition and Examination of Key Elements & Processes." *Simon Frasier University*, wiki.iat.sfu.ca [Accessed April 18, 2017].

Pearlman, Ellen, Stephanie Dinkins, Max Fowler, and Rachel Halberstroh, 2017. "Research Roundtable 1: Creative Collaboration." *Pioneer Works*, https://pioneerworks.org/classes/research-roundtable-1-creative-collaboration/ [Accessed July 29, 2017].

Pearlman, Ellen, and Andy McWilliams. 2015. "Art-A-Hack." *Art-A-Hack*, https://artahack.io/summer-2018/call/ [Accessed December 27, 2017].

Pepperell, Robert. 1995. "The Posthuman Conception of Consciousness: A 10-Point Guide." *Consciousness Reframed III*, http://www.robertpepperell.com/papers/The Posthuman Conception.pdf [Accessed May 29, 2016].

Weiss, Howard M., and Russell Cropanzano. 1996. "Affective Events Theory: A Theoretical Discussion of the Structure, Cause and Consequences of Affective Experiences at Work." *Research in Organizational Behavior* 18 (1), pp. 1–74.

12

DELIVERING HACK EVENTS WITHIN THE ARTS*

Victoria Bradbury and Suzy O'Hara

Influences and Motivation

We began developing hack-style events together in 2014 with 'Thinking Digital Arts // Hack' ('TDA//H'). Victoria is an artist and educator while Suzy is a curator and researcher with an established arts production background. These perspectives have strengthened our partnership, as one of us is always focussed on how the participants think and make from the inside out and the other, from the outside in. Adjusting the hackathon format, originally from a commercial digital model, to work within the arts presents logistical challenges to organisers. This includes modifying components such as competitive aims and needs that are specific to arts participants, strategically planning materials and resources and carefully considering exhibition formats.

What Is a Hack?

Hackathons, which are more commonly referred to as 'hacks,' are a format for the production of collaborative projects. They are intensive events that occur over a short time frame (from a single day to 48 hours), which encourage problem solving and co-making. While hacks originated within the commercial digital sector as an innovation strategy for developers to think creatively

* A longer version of this paper was first printed in: Bradbury, Victoria, and Suzy O'Hara. 2016. "Evaluating Art Hacking Events Through Practice." In *Cultural R>Evolution: Proceedings of the 22nd International Symposium on Electronic Art (ISEA 2016)*, edited by Olli Tapio Leino, 27–35. http://www.isea-archives.org/docs/2016/proceedings/ISEA2016_proceedings.pdf.

about solving technical problems, they represent a style of working that artists have historically used. The 'hackathon' format, however, is considered to have been appropriated by the arts and cultural sectors from the commercial tech sector as a way for artists and organisations to work in an experimental, low-risk way.[1]

Wikipedia has defined hackathons as they are most commonly known within the commercial digital sector, stating,

> A hackathon [...] is an event in which computer programmers and others involved in software development and hardware development, including graphic designers, interface designers and project managers, collaborate intensively on software projects in competition with other teams.[2]

Hacks, however, have been used by cultural institutions, artists and curators as a strategy to generate discourse, collaboration, and as a starting point for new artworks and ideas. These hacks represent anti-disciplinary practices emerging at the intersection of arts, culture, creative digital industries, design and engineering. Hack formats are shaped by the motivations that drive their development. Various agendas reveal the needs of vastly differing communities of professionals including artists, hackers, makers, creative coders, engineers, scientists, data architects, technologists, arts and cultural professionals, venues and business. The hack format is portable and can be effective in commercial, arts and cultural contexts and increasingly, in public service contexts.

Challenges with the Hack Format

From a practitioner's perspective, there can be a number of challenges with the hack format. The rapid nature of the hack can create the impression that art made with technology is 'quick and easy.' While teaching new media to undergraduate students, Victoria has become aware that the public perception of technology making things 'easier' and 'faster' can also make it difficult for students to understand that learning software, hardware and code takes a great deal of work and practice. Textile artist Annet Couwenberg addresses this idea when she states that beginning to use technology in her work '...didn't make things easier...'[3] Digital tools are like any other media that require knowledge and skills to develop alongside of creative and conceptual considerations. While open source code, pre-built software, API's and libraries allow artist-coders to avoid programming everything from scratch, there is still a great deal of work to be done to employ these technologies effectively in projects. These considerations can be particularly acute when projects are generated collaboratively over a short time frame.

In relation to the exhibition of hack outputs, organisers must carefully consider the implications of possible modes of presentation. Facilitators must be sensitive to the professional needs of participants, many of whom work to build their public profiles as artists, both off and online.

This issue of exhibitions after hacks has been notably considered in relation to Art Hack Day[4] events. Art Hack Days are globally peripatetic, driven by a particular theme and held in collaboration with venues and events within the new media art field. Art Hack Day is a non-profit organisation that runs hack events and aims to

> …bridge the gap between art, technology and entrepreneurship with grassroots hackathons and exhibitions that demonstrate the expressive potential of new technology and the power of radical collaboration in art.[5]

The Art Hack Day organisation and the hack format in general received criticism from arts communities for the event format used in the 2014 Transmediale 'Afterglow' Art Hack Day in Berlin, Germany, which involved a 48-hour research and development period during which invited artists would produce work for a 'flash' exhibition. Artist Constant Dullaart publicly declined his invitation to the event, highlighting emerging issues around the real costs of '…experimental innovation…'[6] that utilise commercially based methodologies to generate art for an exhibition context. In an open letter to the 'Afterglow' organisers, Dullaart raised questions around the motivation, agenda, working practices and dissemination strategies that were employed. Dullaart expressed concern around the '…creative corporate…'[7] hack format, time frame and context being created for art production, stating,

> A fast, cost effective, even competitive, corporate way in which a large quantity of approaches can be included, competing with each other, stimulating ridiculous work hours, without any fee or compensation. Stimulating easy and quick solutions to personalise mass produced technology with an artistic flair. After which the work is presented without any chance of contemplation, or for that matter curatorial intervention…[8]

From Dullaart's point of view as an artist, issues of finance, payment and fair working practices need to balance the perceived value of the experience of collaboration. Dullaart's comments highlight that the curatorial decision to host an exhibition following a research and development-focused art hack event can, in fact, be detrimental to the experience of the work produced, the artists involved and the intended audience. The hack format presents potential challenges for organisers, venues and participants when it is migrated from the commercial digital sector into an arts and cultural context.

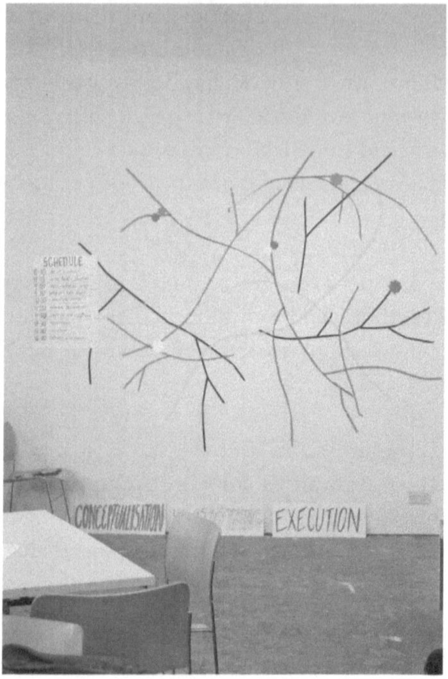

FIGURE 12.1 *Thinking Digital Arts // Hack* preparation for ideation phase. Newcastle, UK, 2014. Photo by Lalya Gaye.

Thinking Digital Arts // Hack

The 'TDA//H'[9] took place on Tuesday 20 May 2014 in Newcastle, UK. The hack was part of the public facing and publicly funded 'Thinking Digital Arts' (TDA) festival, which was devised and curated by Suzy. TDA was curated to run alongside 'Thinking Digital,' a TED-style commercial tech conference that is produced annually at the Sage Gateshead, Newcastle. Now in its third year, TDA began as a practice-led research case study for Suzy's doctoral research project. The programme was designed to test experimental models of creative production that interrogate the impact that collaborations between the arts sector and the commercial digital and creative media industries have on artistic and curatorial modes of practice.

TDA//H sought to establish a dynamic context for hands-on co-creation, experimentation and risk taking and collaborative making between cross-disciplinary DIY makers, artists and designers from across the UK and Europe, as well as conference delegates from the creative digital and technology sectors. As participants were made up of artists and creatives, Suzy invited two practicing new media artists and facilitators, Victoria and Lalya Gaye from Attaya Projects[10] to support the development and lead on the delivery of the day.

Strategies of Organisers

While planning 'TDA//H,' we became aware of the concerns that had been raised by the new media arts community around the recently presented Transmediale 'Afterglow' Art Hack Day. We aimed to be sensitive to these emerging issues as we moved forward with inviting artists and planning the event.

The theme of 'TDA//H' was 'Decentralisation,' inspired by the simultaneous public conversations about the 2014 Scottish Referendum and questions raised about the tendency of UK arts to be London-centric.[11] Newcastle, as the hub of arts and culture in North East of England, was an ideal place to bring participants together to consider this topic through collaborative projects.

Victoria and Lalya made it a central goal to acquire and present quality materials for participants. We did this by using some of the project the budget to purchase electronics, Arduinos, sensors and conductive fabric and thread and also by partnering with New Bridge Project[12] and Maker Space Newcastle.[13] These two venues, a white cube gallery/project space, and a maker space, sit side-by-side on New Bridge Street, a central avenue of the city. The collocation of a gallery space and a community led maker space as neighbouring venues provided a particular context that was appropriate to observe a cross-sector arts hack. Holding the event across these two spaces meant that participants could move between them, drawing upon the resources available in the gallery space (including projectors, monitors and video equipment) and also have access to the maker space's facilities and materials (including a laser cutter, 3D printer, basic wood and electronic tools). Maker Space supported this effort by offering volunteers who participated in the hack while also monitoring and facilitating the use of their space.

Our observation of the conversation around the Transmediale 'Afterglow' Art Hack Day influenced the curatorial decision not to host an exhibition of the work produced during 'TDA//H.' Instead, we announced the public event as an 'informal showcase' of the prototypes that would emerge. However, while the language may have been modified from 'exhibition' to 'informal showcase,' and understood by those working in the field, we found it had little bearing on audience expectations on the evening of the event. The lack of curatorial input into the showcase raised criticism from audiences who, despite the marketing message, still expected a curated exhibition. This underlined the fact that both the language and strategies used in organising art hacking events is still being defined and understood.

Outputs

Because of our emphasis on sculptural materials and equipment, 'TDA//H' resulted in mostly physical and 3D projects. The participants made use of the resources provided and found a balance between play, serious inquiry and final production at the end of the event. A range of projects emerged from the five groups, including a bird flocking algorithm and sculptural prototype, a mapping the city video artwork, an unplayable board game made of dissected laser-cut maps and an interactive installation using web coding and Makey Makey[14] that presented the miles that different foods travel to reach Newcastle food markets.

Strategies of Participants

In order to maximise productivity during the short one-day timeframe, we broke the day into three distinct phases: conceptualisation, prototyping and execution. After the conceptualisation phase, there were short presentations by each group. During this early stage, the groups with more participants from the commercial digital sector had refined their ideas to a project that they would pursue through the prototyping and execution phases. This was mapped out quite clearly during their presentations and the outputs that these groups realised at the end of the day were similar to what they had initiated early in the event.

The groups that consisted of artists presented a vague and broad scope of what they planned to investigate. In the end, these groups showed projects that were further removed and more surprising in relation to what they had presented after the initial conceptualisation phase.

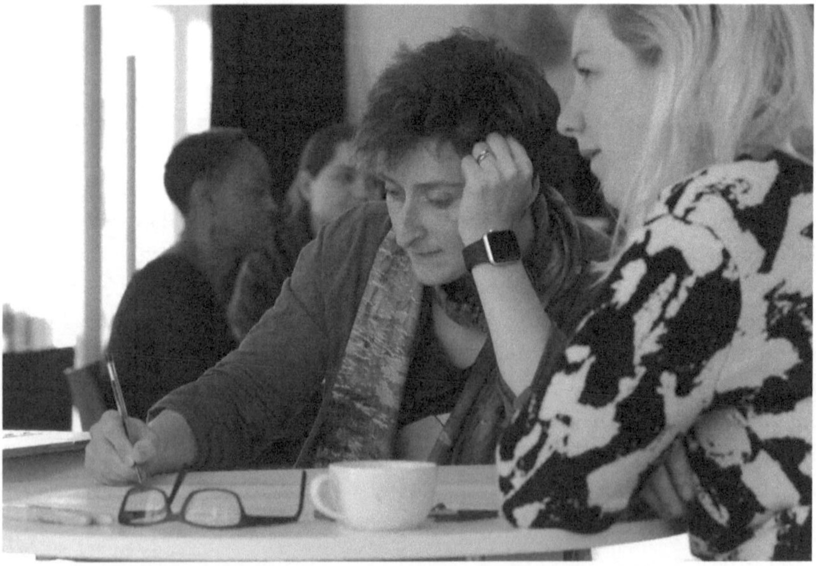

FIGURE 12.2 Participants attending *ReWriting the Hack* at The Core, Science Central, Newcastle upon Tyne, UK, 2015, [left–right] Carmin Karasic, Elizabeth Dobson and Rosa Menkman.

Rewriting the Hack

'Rewriting the Hack' ('RtH') explored female narratives within the theme of 'Industrial and Post-Industrial North East' (England). The female-only hack focussed on issues related to women and tech with an emphasis on past and present histories and archival materials. The two-day event examined the hack format as a site for producing collaborative, interdisciplinary artworks while examining issues of gender diversity in art and tech and in the hack format itself, particularly as it is an increasingly popular model for creative production. 'RtH' was held at The Core, Science Central building in Newcastle on 21–22 November, 2015. Built on an old coal mine, the site is now a tangible space for post-industrial research and digital industries. The hack included participants from around the UK and Europe and was supported through funds from the 'Inhabiting the Hack'[15] research programme.

Strategies of Organisers

'RtH' was organised and delivered by Suzy and artist Shelly Knotts with Victoria in an advisory role during the planning process. The first day of the event began by presenting a live-theatre piece, a monologue by Katherine Beaumont about women's history in the North East of England. Shelly then used the live-coding environment Super Collider[16] to randomise initial groups for brainstorming the key topics of Systemic Obstacles, Structural Inequalities, Voice and Visibility, Revealing Narratives and Feminising Code. Initial subgroups then discussed these themes, followed by short presentations and hanging the topics on the walls of the space. After a break, participants were invited to gravitate toward particular themes and form groups for project conceptualisation, prototyping and execution.

Participants were selected who were librarians, artists, designers, musicians, filmmakers and businesswomen. They gathered to create with archival materials relating to the North East of England's industrial heritage as well as current open data sets representative of the region's post-industrial present. In the weeks leading up to the event, the Mining Institute digitised materials on women's history including several chapters of a book called *We are Women We are Strong* that discusses women's roles in the 1980s' UK mining strikes. This digitisation was a notable result of the hack – the event brought pieces of women's history into online archives that hadn't previously been there.

Outputs

Three groups formed from the 13 hack participants. One focused on a single collaborative diorama project that activated mining history data called 'Haway the Lasses.' Two larger groups worked in more divergent ways, creating a variety of projects around a single topic. These included a series of explorations of business models in relation to gender and a series of projects examining women's inclusion or exclusion in histories.

Strategies of Participants

The 'Haway the Lasses' group were highly focused, conceptualised their project quickly, then moved forward with its development in a linear way. Their diorama examined the domestic side of mining history, drawing quotes from women who were miners' wives, mining statistics and key events that happened when mining was most active in Britain (such as during World War I). The diorama included a timeline with activated LEDs and a pickaxe that moved based upon shifting data.

The business and gender group created a publication called '29 Things to Do at Work' and an embroidered tablecloth that questioned business etiquette for use as a centrepiece at business meetings. They also wrote a script that examined biographies of 90 male and 30 female professors from Northumbria University, making a quiz about words that were used in the biographies and creating an award in the form of a power-biography.

The group considering women's narratives challenged the authority of histories with two key outputs. The first was a loudspeaker that projected male or female voices reading stories from 'We Are Women; We Are Strong.' The voice changed to male or female based on proximity of a viewer. Another member of this group considered what would happen if histories were altered and wrote a script for removing men from the operating system of her computer.

An informal public showcase was held at the end of the second day of 'RtH.' Because this showcase was not part of a larger cultural event or conference, it remained intimate, including the participants and only a handful of visitors. This allowed time for groups to discuss the future of the projects created during the event and served as a buffer between a period of meeting and intense collaboration and finally parting ways.

Outcomes of Hack Research

Presenting hack events in arts contexts transfers a format from a realm that is interested in problem solving into a realm that tends to ask questions rather than answering them. When the hackathon is modified for the arts, there are a number of issues that must be considered, particularly in relation to participants. The hack format aligns closely with methods that are already practiced within the arts, including interdisciplinary practices, collaboration, creating quick prototypes, brainstorming and thinking through action. While these similarities exist, there are also many differences, particularly in relation to the aims of artist-participants in non-commercial contexts. Organisers must be open to and indeed support the idea that art hack participants will pursue goals during the event that will further their own creative and professional agendas; the exploration of new ideas is a key currency for the artists involved.

When the hack format was originally used within the tech sector, it was focussed on the goal of writing and developing code for commercial applications. Writing code and creating programmatic artworks is only one of many outcomes and strategies that are undertaken in art hacks. Artists are accustomed to thinking laterally and applying art-making methodologies across a variety of media. During an art hack, this occurs through experimentation with media and materials as part of the phases of project conceptualisation, prototyping and execution. Art hacking events, therefore, do not necessarily result in technology-based art projects, but rather, a wide range of results can be possible. Some of these results utilise technology while others do not. Participants often ignore technological media altogether and work with analogue materials or performance. This can be an even more effective use of their time, circumventing the possibility that learning a new technical skill will encumber the development of new work.

Organisational models that are presented by facilitators will have a major impact on the kinds of projects that result from art hacking events. How these are shaped depends largely upon the interests, skill sets and backgrounds of the organisers. We have tested different modifications of the hack format across our projects and they have produced a variety of results. An example of this is seen in 'TDA//H.' This event resulted in many physical computing and sculptural projects that were made possible by the materials and resources available in the New Bridge Project and Maker Space venues. Groups were able to use materials such as laser cut cardboard, Makey Makey and video monitors in their final presentations.

The mode of selecting participants (including invited guests, open call curated selection or open invitation) will be reflected in event outcomes. Diverse participants, when grouped, will have divergent expectations, and hack results will therefore differ. We have seen this across our projects which involved participants ranging from artists, journalists, designers, archivists, librarians, filmmakers and web programmers. The combinations that arise in groups, whether by chance or by design, will affect the methods used to create projects and thus the outcomes of the event.

The structure of a hack can be compared to pedagogical models employed in art universities, particularly in foundations programs in which students are led to plan and implement projects quickly while simultaneously expanding their ideas and uses of materials. When the competitive format and the pressures to exhibit work are removed from the commercial hackathon format, we have found that participants in art hacking events are able to create more freely and push the bounds of their practice and usual modes of working.

Organisers of hack events presented within the arts must be aware of the types of considerations that are unique to artist-participants. Outcomes of art hacking events will be both tangible and intangible. These include physical artworks, virtual artworks, ideas, relationships and collaborations. It would be impossible to monetise or place a concrete value on these types of results. It is important that participants who work collaboratively in art hacking contexts discuss intellectual property before the end of the event. If individuals decide to move forward with a project, all group members will be aware of what kind of credit or remuneration was discussed.

Issues of intellectual property and participant visibility extend to the fact that artists labour throughout their careers to build a personal brand. They usually serve as their own marketing department, creating a website, archiving imagery of their practice and acting as the social media arm of their own identity and career. This raises questions as to whether an artist wants to link his or her name with a project that has been created quickly and collaboratively and is then showcased immediately either in an exhibition format, on social media, on a project website or in articles about the event.

Conclusion

Evolving strategies employed in hack events presented within the arts have modified the commercial tech hackathon model. Organisations, facilitators and institutions have pursued a wide range of models for planning and implementing these events. Some follow the tech hack model more closely than others.

In planning and delivering the case study projects, we have worked to de-emphasise the original competitive nature of hacks as well as the need for a public exhibition of hack results. Our attention has been focussed on deriving value from collaboration, idea generation and networking. A key ethos of this hack model emphasises the idea that artists tend to raise questions and create problems rather than answering questions and solving problems. Raising questions can create insight and lead to new ways of thinking that can be considered a valuable type of innovation.

Notes

1 Gunatillake, Rohan. 2012. "The Rise of the Hack Day and What It Means for the Arts." *The Guardian*, February 28, 2012, sec. Culture professionals network. https://www.theguardian.com/culture-professionals-network/culture-professionals-blog/2012/feb/28/hackday-rise-significance-arts-culture [Accessed June 15, 2015].
2 Wikipedia, "Hackathon (2015)", https://en.wikipedia.org/wiki/Hackathon [Accessed November 24, 2015].
3 Joni Seidenstein, Artcollisions, 2015. "The Skeleton and The Skin: Annet Couwenberg", https://artcollisions.wordpress.com/2015/02/23/the-skeleton-and-the-skin-annet-couwenberg/ [Accessed November 24, 2015].
4 See: http://www.arthackday.net

5 Art Hack Day, 2015. "Art Hack Day", http://www.arthackday.net/about, [Accessed November 25, 2015].
6 atractivoquenobello, "An interview with Constant Dullaart (2014)", http://www.aqnb.com/2014/03/12/an-interview-with-constant-dullaart/ [Accessed June 15, 2015].
7 Ibid.
8 Ibid.
9 http://blurringartandlife.com/vb/thinkingdigitalarts_hack.html
10 http://attayaprojects.com
11 http://www.hannahfestival.com
12 http://thenewbridgeproject.com
13 www.makerspace.org.uk/
14 http://makeymakey.com
15 http://inhabitingthehack.github.io/about/
16 http://supercollider.github.io

Bibliography

3beards, 2015. "Growing the Tech Community", http://3-beards.com [Accessed June 15, 2015].

Bradbury, Victoria, 2015. "The Performativity of Code in Participatory New Media Artworks", Ph.D. diss., University of Sunderland, 138.

British Council, 2014. "Culture Shift", http://creativeconomy.britishcouncil.org/projects/culture-shift/ [Accessed June 15, 2015].

Inhabitingthehack.github.io, 2015. "Inhabiting the Hack by inhabitingthehack", http://inhabitingthehack.github.io [Accessed July 22, 2015].

13

WHERE DO WE WORK? THINGS WE CHAT ABOUT WHILE SORTING SIM-CARDS

A Conversation with Constant Dullaart

Victoria Bradbury

I visited Constant at the International Studio & Curatorial Program (ISCP) in Brooklyn on a September morning in 2017. Walking through East Williamsburg, I stopped to grab a coffee and some croissants to share. A few minutes after settling on the front steps of the ISCP, Constant arrived by bike in an Oxford shirt ready to work. Entering the building and walking upstairs, we found Constant's studio large, brightly lit and teeming with projects, equipment and images that the Dutch artist had been working with over the nine months that he was based in New York. A mountain of SIM-cards covered one of the tables. Constant invites artists to sort the cards when they visit, an exchange of labor for chat. As I began to carefully sort orange, green, blue and white cards into piles, we talked about practice in context, where we as artists find ourselves working and how the spaces and situations we place ourselves in affect the results of our making.

VB: One thing we have in common is that we have both worked in China in different contexts. In 2012 – you were based at the OCAT residency in Shenzhen and I was in Shanghai working out of Xinchejian Hackerspace. What was the OCAT residency like?

CD: OCAT is large – I had the sense that state money was used to set up a relevant international arts organization, but in terms of my presence there, I felt at times almost like a décor piece – I was flown in, I was documented everywhere, like if I went to an exhibition, I was photographed looking at everything. I felt like this creature – an overweight, kind of awkward white person looking at art. At the same time, I felt very much like I came in with my Western morals as an artist to form an opinion about how other cultures function. For me as an artist, though, I felt that I needed to engage with local culture.

For my practice, I wanted to find out what the tech and fabrication industries were like in Shenzhen and to visit a factory. When I was successful at accessing a factory, I suddenly became very aware of my position of privilege but also of a certain bias that I had coming to Shenzhen as a Westerner. I saw that the people making the boards for my future project wore a bracelet that tied them to their workstations. I felt horrified because I thought they were shackled to the production line! I asked the factory owner about this, and they explained that it was a safety measure – because they were working with electronics – this device managed the static electricity. That experience made me aware of the fact that I came in with a prejudiced line of thinking that I would understand the working conditions in Chinese factories.

This was quite a remarkable moment in which I felt I didn't actually want to be so judgmental and black and white over these kinds of things in relation to how the DullTech™ player was made. It made me want to redefine my own relationship with technology, and figure out how I could be respectful, but also create a project that would comment on these complexities and differences. So that is what this residency offered, a situation of 'cultural adversity' in which to envision projects.

VB: The first time I worked in China was in 2008 as artist-in-residence at Imagine Gallery, Beijing. This was a traditional arts organization –the setup of my residency would have been ideal for someone working in painting or drawing. There was a large studio, but it was tailored as a space to work with materials such as paint or clay, and it had unreliable WiFi (it *was* 2008). Looking at the short time I had, about four weeks, I started to compare the Pizza Huts I passed on the street in China to those I was accustomed to in the US. The Beijing Pizza Huts were quite fancy and offered a nicer dining experience. I was interested in this difference and ended up finding a card that listed the address of every Pizza Hut in Beijing. At this moment before smart phones, a card like this was incredibly useful, but I doubt it would be produced today.

CD: Where did you find this card?

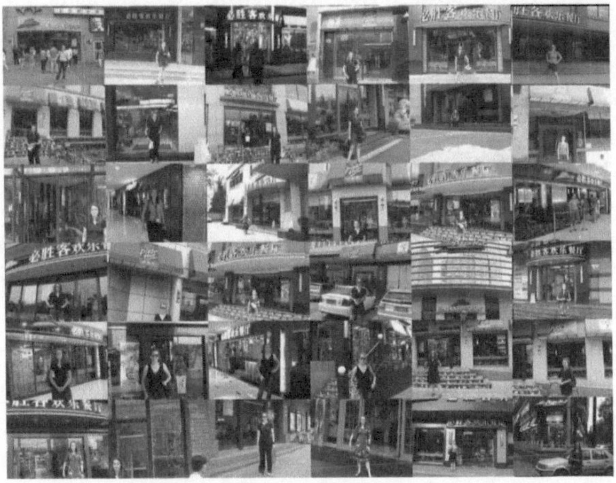

FIGURE 13.1 *37 Red Hats*, Victoria Bradbury, 2008.

VB: I had already decided I wanted to visit every Pizza Hut in Beijing, but I had no list of them. So, I started going to a few of the restaurants, and at one of the first, the hostess gave me the card with every location address! These addresses, of course, were listed in Chinese characters, so I was personally unable to read them. The residency gallery coordinator helped me to plot the addresses from the card onto an English-language paper map of the city. Beijing is massive, which made our plotted points mere approximations. Each day, I propelled myself to an area by a combination of foot, bus, bike or subway. When I arrived, I would start asking people where to find the address I was looking for on the card. Beijingers would point me in a direction, and I would walk until I found the Pizza Hut, or *Bìshèng Kè* (必胜客). I would usually have to ask four or five people before I found each place. When I arrived, I would ask a passer-by to take a photograph of me in front of the restaurant, so I ended up with a series of photographs of myself in front of every Pizza Hut that was in Beijing in the summer of 2008. I was at the same time a flâneur and an absurdist tourist with an extreme mission, navigating the city with the address card as my score. The resulting photo database became a record of maneuvering through the city as a foreigner with Pizza Huts as my guide. I was hoping to understand more about the relationship between China and the US by centering commerce, food, tourism and culture in this way.

CD: And no one would question why you would go the Pizza Hut? What is it called in Chinese?

VB: Bìshèng Kè. But, yes, no one seemed surprised that I was looking for the Pizza Huts, they were usually happy to help me find one if it was nearby. Later, when I was in China in 2012, I was working on practical projects related to my PhD research. I was based in Shanghai for a period of time because my husband was working at a glass studio there. Because I didn't have a dedicated studio, I started to work at Xinchejian Hackerspace, which is the first maker/hacker space in China. Working there afforded me opportunities because I was able to meet all types of people, including artists, designers and researchers in the community. This became a social and professional resource that allowed me to talk about my work and learn about what others were doing.

The project I ended up working on began as an idea to make a backpack-style translation device. It would be my laptop on my back, in a body suit and it would translate English to Chinese live on the spot as I moved around the city.

Now of course the technological issues that came up with this immediately were … that there was no Internet connection if you were out and about (I was going to tie it into Google Translate)… and this was 2012, so still early days for Google Translate. What I ended up doing was using Processing to code a crude translation device that worked with a

Wi-Fi connection. I incorporated a pushbutton and a microphone inset in a custom Perspex box. I was then able to use it to have conversations with people at Xinchejian.

CD: And the people at the maker space, were they artists?

VB: At Xinchejian, many people were working as technologists but had studied art and design or programming. Some were working on start-up ideas and there were also hobbyists and students there to meet other people interested in technology.

CD: I was wondering because the vibe I have, especially here in New York now, and in other maker spaces too, is that they are filled with people who are trying to make the next smart dildo or the next networked pottery-water-plant thing. I always feel like everyone has their elevator pitch ready to be successful by way of Arduino.

VB: That's something I have found difficult about working on art projects in maker spaces … someone always comes up and asks what you are working on … and to have this kind of elevator pitch ready for your project is odd for an artist because the project is still in development and you're not necessarily up for talking about what's going on because it's about this messy space as you are shaping what your project is going to be.

CD: And in a functional way … trying to convince people of the validity of what you are doing, because sometimes you are doing research and you don't know if it's valid or not, you just have to figure it out, or run through a couple of details. It might not even relate. I spend a lot of time just watching demo videos to see how something is done as I think of how to misappropriate that knowledge or use it awkwardly or misapply a way of doing something. If someone sees that I'm just watching YouTube videos forever … that is difficult to explain … no I'm just watching this strange thing where somebody tried to build a speaker, I think it's going to be relevant … but not immediately.

VB: Right. There's a lot of research time.

CD: The question also becomes, who is funding the maker space? In these spaces, I'm often wondering, did this 3D printer come from a company that is trying to increase filament sales? Like how software companies give student discounts because they want students to continue to use their software after graduating. I'm also thinking it gets close to the idea of incubators where you get the sense that the person running the incubator wants to be culturally affiliated with the participants. They can choose artists with a critical perspective but they can always host them alongside a commercially viable project. Even if they hate each other, the incubator always goes away with the prize no matter what happens. Like if the critique is functional and the product takes off, then they have been a successful host. The artist becomes an ornament to make the commercial context look cooler…

VB: The artist can become an ornament in the 'making' context, but how is that different from the artist being invited to an exhibition by a curator? There's some cultural attaché there for the curator and venue as well.

CD: It's true the curator can get more professional traction or more social agency afterwards than the artists that were in their show… But if I show work, it is (mostly) not in-progress, it's the work that is manifest within the conversation of the show. I see an exhibition as a conversation and my work is the argument within it. The process to get to that argument takes place in the studio, but is inherently not the argument itself. Publicly hosting this process by a maker space, or incubator says the host is the enabler of my agency. And with an exhibition, I prefer to show work within a self-chosen context, and not to say it was enabled or initiated by Hewlett Packard or Microsoft who had some extra office space and wanted to complement their corporate image with the cultural capital of critique.

Sometimes I find it troublesome to get into situations where I feel like an artistic ornament. You want to have an artistic practice that sits slightly outside of society to be able to comment on certain structures that you can easily become dependent on. Like social media to amplify your efforts, or crowdfunding. You don't want to bite the hand that feeds you, but it's embarrassing to sell t-shirts and tote bags when you want to criticize something like, let's say, capitalism for instance.

VB: So there's the artist as cultural capital and the artist as jester. If the artist is invited to a panel at a tech company, and the artist is there with a different perspective on the topic at hand but they are also looked at as silly or not taken as seriously because they're out of their own context…

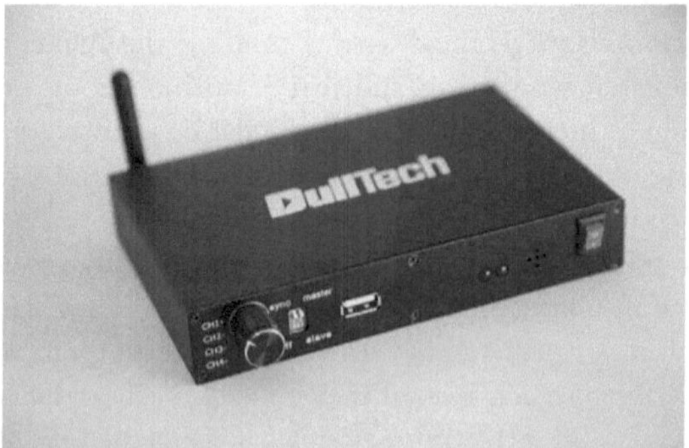

FIGURE 13.2 Constant Dullaart, *DullTech™ player*, 2014.

CD: … well, that's also what was interesting about starting a company. With DullTech™, I created my own umbrella under which I could be artist-in-residence of my own company … but also this company had its own type of agency. If I was asked to be somewhere for DullTech™, my words carried differently. Like, there is actually a company that is commercially viable? And if it is commercially viable, there's another meaning to your words. This became very interesting because I could say certain things and be critical of stuff, and because it was commercially viable, the critique became even more valid.

CD: In Berlin, we went to visit a bunch of co-working spaces. Some of them have expensive espresso machines, and some have slightly less expensive espresso machines next to a 3D printer, as if 3D printing was just an extra thing you could do, quick prototyping in the co-working space. In the end, these were still commercial ventures where you don't have to build walls, just put up desks instead of having people rent out an office. Like maker spaces, I wonder, what is the benefit? Is it important for the funder that everyone understands the 3D printing revolution so you don't have to buy things and you can just print them out? In short, do we know why the space is there, is the incentive transparent?

VB: Right, for the 'greater good.'

CD: It's hovering between the same thing with residencies, turning back to working in China, I felt when I was being photographed that I was representing Western art culture that had been flown into China (my participation had been co-funded by the Dutch Mondriaan Fund). At the same time, it was strange because I was in a state-run residency so I couldn't talk about human rights or what the organizers thought about open Internet access or VPNs, Facebook, things that were illegal. It is interesting to consider who is taking the cultural credit…

VB: And then tying that back to setting up your own company and becoming an artist-in-residence *there* becomes the ultimate artistic gesture…

CD: Developing Dulltech™, it was interesting to be embedded in this production chain – in the West, we all have hardware that comes from China – most people work on their laptops without wondering where they came from. What's the actual geopolitical reality that's behind that screen?

I have learned things about business, but there is also the socio-political aspect as I was going into a cheap labor market and criticizing the process while doing it at the same time. It felt strange to continue that. Because the product in the end became successful (it's been sold out for over a year and I still get a lot of requests from people wanting to buy it), but it was interesting because I really didn't feel like continuing that process. Currently I am trying to develop new ways of local production, and combining them with education and critical exchange, kind of like we are sorting SIM cards right now.

So, I decided part of the labor needed to happen locally, where I can control it more like a performance. We decided to develop that idea because that would be a model where I could hire a person in the artist community to package the players, print t-shirts, rather than producing them in China. For example, there's a distribution warehouse in Germany where we can build the players in an art institution nearby, and bring together students and artists who can assemble these boxes and talk about their work at the same time.

VB: So that almost becomes a hack event because you're bringing everyone together...

CD: Yeah, but everybody gets paid. It becomes much more of a community thing. Instead of exporting all the production to China and shipping it everywhere in the world ... why not do the assembling somewhere else?

VB: Do it locally, where it is going to be exhibited and sold.

CD: It is still a viable product, but the discourse becomes slightly more inclusive as it ties back into this commentary where you aren't outsourcing the labor anymore. I think it's more interesting to figure out where the use of labor sits within the work or within the artistic process, if it's other people's labor, that becomes part of the material. It becomes about whose labor is applied so that's not just arbitrary.

You were saying previously how as an artist you are trying to challenge technological adversity all the time. And how you're in this strange position always trying to figure out a new technology or media with specific qualities.

VB: And that is why many of us are drawn to technology as a medium. We like always trying to figure out how things work and we want the work to be in a place of technological discovery. But in a maker space, everyone is there with different motivations. You have hobbyists, aspiring start-ups and sometimes you have artists ... and the issue becomes about labor and trading labor ... is the hobbyist always the one who has to show how the 3D printer works so the entrepreneur or artist can use it?

CD: Do you think that area should be a concern in maker spaces?

VB: Not necessarily because as an artist, perhaps I'm not meant to be there.

CD: You should go there only if you want to be a craftsman on the 3D printer?

VB: That's the question – as an artist, I assume I'm a maker, or one could say that. Do I belong in a maker space or not?

CD: It's also strange because the maker space suggests an idea, and the hacker space to a certain extent too, into reversing the normal flow of consumerism. Like you would say, I could just buy something in the store, but no, I'm going to make it – make my own version of it. And that often turns into, will this be successful? Like you made your own version of this, that is genius, you have a product. Like the faint remnants of the American Dream where somebody can make their own body scrub. I remember my wife made a body scrub and it was like, 'that's brilliant body scrub, you should start a business,' the typical American Dream. Like you should start a business, you should call it this, print a really authentic label, put it in stores and that's the big dream … that's how you can pay your rent. Everything can become a transaction.

But the weird thing is that this maker space or the hack thing, it's not necessarily just a critical perspective. It is not productive enough if you just want to deconstruct it or build a theoretical framework around it. It is like I am engaging with the process or manipulating it so it can be a more successful version of something. It conflicts with the autonomy of artistic research, which should be just outside the realm of necessary relevance.

VB: But there's also the idea that we're going to hang out in a maker space to figure out how something works, which also sits outside of my interests. If I'm going to hang out and figure out how things work, it has to be part of a larger project or have a pedagogical aim.

CD: And so, it feels like these kinds of institutions are social alibis that need a final argument – What did you do? What are you working on? And I can do it with my work now too, like I'm the person who discovered the first photo that was Photoshopped, I ran a Facebook army of 13,000 artificial accounts that discussed the value of news propaganda online and now I'm building a monument (with these SIM-cards) for these deceased propaganda soldiers and I'm running my own technology company that is ironic but successful. But when I am thinking or doing my research, I'm not thinking in sentences like this – the ideas aren't verbalized in soundbites.

Like that technological adversity, sometimes that struggle needs a fine, fertile ground. Like you can struggle on the level of how to find an on-switch of a 3D printer, or you can find the level of how to apply the machine to a certain detail.

VB: Or how to locate something in a city where you don't speak the language?

CD: I always think there should be a type of independence of the artist to have a cultural critique, or to be able to choose a position and reflect on something without being surrounded by affectivity or solutionism.

VB: Affectivity and solutionism are some of the baggage that come along with joining a group with a corporate agenda, which art shouldn't necessarily fit under.

CD: But that's why it's great. DullTech™ can just be a company that doesn't have a product for a year. Well, we did have a product, it was just a website that can make images more boring, more tawdry, so we could make that, but there's no commercial product right now. We'll probably have it in a few months, but I don't know exactly how it will take shape. If I was making this in an incubator, I would have been pushed to make more so I could validate my efforts without too much time to reflect.

Leaving Constant with many SIM-cards unsorted, I walked around the corner past Transfer Gallery, where I had seen Rosa Menkman's Behind White Shadow's VR piece the day before, and strolled back past car repair shops and bodegas. I reflected on our conversation, thinking about the many projects Constant has going on at once and the messiness that mixing technology, materials and spaces brings to my own chaotic process.

14

ART HACK DAY

Olof Mathé

Introduction

As the name implies, *Art Hack Day* lives at the intersection of the art and tech communities. Both 'art' and 'hacking' are multi-faceted and rich terms that defy easy definition. Questions like 'what is art?' or 'what is hacking?' tend to elicit vivid discussion. 'Art' is more than the application of human creativity, and 'hacking' has wider connotations than gaining unauthorized access to computer systems. The meaning of both terms is hotly debated and constantly evolving.[1] *Art Hack Day* doesn't try to contribute to their definition. Rather, *Art Hack Day* is an event for doers, and while we don't reject theory, we emphasize the process of making as a way of articulating thought and bringing it to life.

The process adopted at *Art Hack Day* events, and the works created, are living testaments that we can think about both art and technology differently. We want to demonstrate the expressive potential of technology, whereas hackathon culture is all too often utilitarian, emphasizing virtuosity over insight and meaning. We want to bring radical collaboration to the art world, to explore what teamwork, prototyping and open source approaches can contribute to art.

Art Hack Day tries to bring both of these traditions closer together by inviting both artists and hackers to see what collaboration can create. Later in this chapter, we'll look at two different works created in the course of our multiple events: "Iconoclashes" and "Glacier Pop" to show the ways in which we've achieved our goals. These works demonstrate that authorship is a diffuse notion and that people from all walks of life can create art. They show how different artistic practices can engage a broader, more diverse public. They also highlight ways in which art can thrive outside of the institutional art world, the white cube gallery and the art market.

First, however, let's explore the beginnings of the event.

How Did Art Hack Day Begin?

Inspiration often comes from unlikely sources. In my case it was triggered by a commercial hackathon called *Music Hack Day*[2] and a conference called CES.[3] I was also lonely in my art practice and frustrated with how hackathons were run.

At the time, most hackathons perpetuated stereotypes about tech culture and were commercially driven events, spearheaded by the latest crop of tech companies. Ultimately, many were thinly veiled recruiting events, or advertisements for their latest developer APIs.[4] To me, the spirit of hacking seemed lost from these events. While everyone was friendly, there was little collaboration. They were frequently competitive: prizes were given out by the sponsoring tech companies. The events ended with participants "demoing" or showcasing the functionality of their projects onstage, thereby favoring virtuosity over insight, and assuming that attendees wouldn't be able to understand the projects unless they were part of a presentation.

Unsurprisingly, participants at these hackathons were overwhelmingly white and male. Having participated in several of these events, I asked myself: how could tech culture evolve to be more inclusive? Could we take a more critical stance? Could the hackathon format itself evolve to nurture a more intersectional attitude?

These questions stuck with me. Then I attended *Music Hack Day*, sponsored by companies like Spotify and SoundCloud. Since music is an art form, this event raised the question of whether the format could be adapted to the art world itself. The idea was intriguing. Any project I had done at a commercial hackathon had aged poorly. At best, the projects were interesting snapshots of the current state of technology, but they lacked lasting impact. In contrast, all the technology-based art projects I had done spoke to me years after their completion. They had longevity, even when the underlying technology was outdated, since the impact of art transcends the particular technology it is made with.

Around the time of *Music Hack Day*, I was working at Skype. It was a good fit for my interests: I was keen to explore the expressive potential of video calling, and how technology can intermediate and form new types of communication.

Skype soon sent me to attend CES, perhaps the world's largest technology conference. There I saw a 'remote presence robot,'[5] basically two wheels on an axle, supporting a tall pole which held a camera and screen. The business idea was that companies would buy the devices and have remote colleagues dial in to them, so they could be present remotely, while still having physical presence through the device. Balancing upright on just two wheels, they seemed over-engineered: Why not just put an iPad on a remote-controlled car, and have someone dial in on Skype? The total cost and development time of such a setup would dwarf what had gone into the elaborate remote presence robot I had just seen.

This remote presence robot merited some kind of artistic reply. At the next technology conference I attended, South by Southwest[6] in 2011, I decided to do a performance to interrogate notions of remote presence. Presenting work at a commercially focused conference felt like a more relevant context than an art gallery, since that's where the audience I wanted to engage was.

The performance consisted of me walking around the conference wearing a mask I had made, with two iPhones as 'eyes.' People could dial into my 'eyes' remotely, see what I saw and interact and communicate with other conference attendees, but without controlling my movements. As a variation on the theme, I turned off video calling and turned on recording on the front-facing iPhone cameras, so that the 'eyes' became mirrors to anyone who looked into them and engaged with me. The mask looked polished, although it was fundamentally 'hacky.' It was built out of a waste basket turned upside

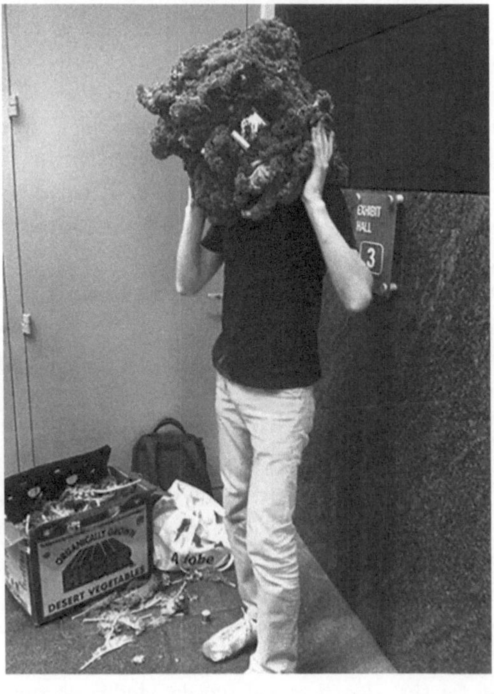

FIGURE 14.1 The remote presence performance at South by Southwest, 2011. Credit: Olof Mathé.

down, completely covered in kale, to give it an organic feel to contrast with the slick iPhone eyes.

The performance itself was gratifying. Conference attendees flocked around me to engage with the work and the people dialing in remotely, and asking me questions ranging from the 'how' to the 'why.'

After the performance, I realized that it hadn't taken more than 24 hours to complete all-in-all, from conception to execution, as if it had been part of a hackathon. My only regret was that I had done the work alone rather than with peers. The thought grew on me: could we organize an event to do similar projects, with peers, under similar time constraints, outside of the traditional confines of the art world?

As soon as I got home, I messaged the New York Hacker list, to see if anyone wanted to organize a hackathon where the goal was to create art. David Huerta and Paul Christophe responded, and in turn reached out to others in the community, including Lindsay Howard, Sofya Yuditskaya and Sam Hart. Igal Nassima, who ran the gallery 319 Scholes in Brooklyn, kindly agreed to host us and the first event got off the ground.

Recent experimentation around art and hacking at that time included "speedshows" by Aram Bartholl,[7] in which all the computers in a public cyber-cafe were taken over to show browser-based internet art. Rafael Rozendaal had launched "BYOB" (Bring Your Own Beamer),[8] in which artists would do just that, and bring their own projector to project their latest works on the walls of a gallery for one evening. Collectives like FAT[9] were making art that interrogated technology, predominantly using similar 'speed'-based methods. However, if you weren't an artist, you were unlikely to participate, and the exhibits didn't have much reach outside of the boundaries of New Media Art practice.

Art Hack Day seemed like a natural evolution of this nascent type of practice, perhaps even a way to democratize it. The first event was themed 'Hackers as Artists,' and over 30 artworks were created and exhibited at the 319 Scholes gallery in January 2012. Over 500 people showed up for the one-night opening, and new friendships were formed among the 55 participants.

Initially, there were no plans for the event to happen again. However, once I returned home to San Francisco, Josette Melchor, Barry Threw and Mat Dryhurst, from Gray Area,[10] expressed an interest in hosting an *Art Hack Day* and in December that same year the second *Art Hack Day* took place, this time in an abandoned retail store on Market Street in the middle of the city.

In preparation for the second event, we decided to formalize its organization, and soon the event crossed the Atlantic to cities like Berlin, Paris and Stockholm.

What Is Art Hack Day?

Art Hack Day is a non-profit organization, and all events are financed by smaller cash or in-kind donations from local commercial and non-profit sponsors and partners. This support is used for basic costs of the event, including catering, infrastructural and material costs (there are no fees for curation, organization and delivery, a subject we'll expand upon later in this section). This portfolio funding approach was not by design, it just so happened that most companies weren't willing to commit larger amounts of sponsorship to an art event that was still at an early stage in developing its profile and audience. The portfolio funding approach actually turned out to be a blessing in disguise, since, with smaller donations, each sponsor expected less in terms of input and their footprint on the event was limited to logos on the website and exhibit program.

Art Hack Day organizers tend to be part of the local 'art and tech' communities and often have an art-based venue in which to host the event, eliminating the cost of rent. Organizers begin planning the event by inviting a core and diverse group of participants who will be able to relate to the format of the event, before opening registration to the public. Participants come from different backgrounds and communities and have different practices, different areas of expertise and varying degrees of professional accomplishment; many don't consider themselves artists or part of the traditional 'art world.' Participants have included not only digital artists and practitioners but also startup enthusiasts, sculptors, dancers, hackers, engineers, photographers and graphic designers too. By design, we strive to have roughly 50 participants. That way, the event is big enough to support a good diversity of backgrounds and practices and ensures that participants always have someone new to talk to and bounce ideas off of, or ask for help. A larger event can cultivate a sense of anonymity that is antithetical to the event and the sense of community it fosters.

Events typically start on a Thursday evening, ending in an exhibition opening on the Saturday night. The event kicks off with a meet and greet in the host venue (this can be as traditional as a formal gallery, or as ad hoc as an abandoned retail space) where participants meet for the first time. Organizers introduce the format of the event and explain what participants should expect during the coming 48 hours. This is also an occasion for participants to introduce themselves to their peers and share project ideas. In order to aid this process, we have also opened up an email list or chatroom for participants to share and elaborate on ideas prior to each event. Once the introductions are over, we serve dinner, so participants can get to know each other informally and form project teams. The Friday is spent working on projects and occasionally, teams work through the night so their work is ready for the opening.

On Saturday afternoon, workspaces are dismantled, the space is cleared and a complete art exhibition takes form. The aesthetic of the exhibit is often one of a white cube gallery, not a busy makerspace, as you might expect. After all, the event has a specific goal, and that is to create new artwork. We found it helpful to legitimize the type of artwork created by making sure that the opening feels professional and polished. The 'flash' exhibition is open to the public for one evening only. There are no 'demos,' as you'd have at a traditional hackathon, the works are allowed to speak for themselves and performances are not uncommon. It's worth emphasizing that none of the works shown are 'half-baked' or in a state of elaboration, the exhibit looks like a professional gallery show. The following day, the exhibit is dismantled and the artworks are collected by exhibitors. Both the hackathon and the exhibit are documented by the organizers, using both photography and video that can be found on our website.[11]

It's worth noting that while the event normally runs from Thursday to Saturday, this has precluded certain people from participating, even if many people go to work for a half-day on Friday. In some cities, to be more inclusive, we had the event run from Friday through Sunday, although the atmosphere of a Sunday opening is often less festive and well attended than that of a Saturday.

It is important to note that no one gets paid to participate in an *Art Hack Day* event as we felt doing so would be antithetical to the social and community-led spirit of the event, which is a hackathon after all. This is by no means an unproblematic stance. The methodology of *Art Hack Day* responds differently in different contexts. When the event takes place in an abandoned retail space, unaffiliated with any official art institution, it's easier to see how *Art Hack Day* creates an ephemeral world of artistic play.

In the cases where *Art Hack Day* was part of a festival, that became harder, and the event acquired a different meaning due to this new context. For example, when *Art Hack Day* was the opening exhibit of the new media art festival Transmediale in Berlin 2014,[12] the exhibit was up for five days, rather than one night, and legitimate questions were raised: was this a way for institutions to 'outsource' the curation and creation of exhibits? Were artists being exploited to work for free?

This highlighted the tension of working at the intersection of art and technology. Whereas technologists typically have lucrative jobs and hacking and open source contributions often serve to increase their reputation, which in turn helps them land even more attractive jobs, the contrary cannot be said of artists. For artists, their artwork is, or at least should be, their main source of income, and the methodology espoused by *Art Hack Day* can be seen as undermining their means of subsistence. Ensuring that the exhibit just stays up for one night has been one way to meet these concerns, since it puts the work in a new context and re-emphasizes the experimentation and risk-taking inherent in the event.

Themes and Artworks

A unifying framework or theme has proven useful in bringing diverse communities with different practices together. Therefore, each *Art Hack Day* has had a theme that we felt had artistic, political and technological resonance. The theme serves as a creative launch-pad for participants and doubles as an evocative curatorial statement. When choosing the theme, we ask ourselves what is topical, in particular given the venue and city, and also, what will be a good 'soundbite.' It is not a coincidence that the themes often sound like movie titles.

The works created at *Art Hack Day* range from the political to the playful, from the technically daring to old-school analog, from the visually striking to the conceptual, and from the sublime to the simplest prank. The diversity of work gives the one-night exhibit its vitality and breadth, and the theme serves as a unifying framework to guide visitors to the exhibit.

Our second theme "God Mode" (NYC 2013),[13] paid homage to the cheat code in video games making players unbeatable. We investigated notions of invincibility, omnipotence and the consequences of backdoor access. We interrogated the near religious reverence granted the most accomplished artists and hackers. In a stroke of divine intervention, Pope Benedict XVI resigned on day of the opening.

Iconoclashes[14] was created by artist Clement Valla, photographer Erik Berglin and developer Jonathan Dahan, three participants who had never worked together previously. In line with the theme 'God mode,' *Iconoclashes* created new, speculative and iconoclastic (pun intended) icons by mashing up images found in the digital image collection of the Metropolitan Museum of Art.

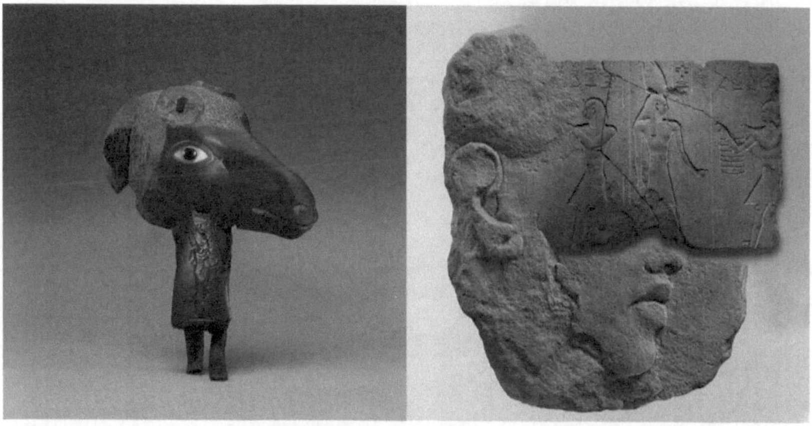

FIGURE 14.2 *Iconoclashes* from Art Hack Day God Mode. NYC (2013). Credit: Art Hack Day.

The group searched for images tagged with 'God' or 'Religion' and fed the images drawn from different time periods, cultures and religions, into Photoshop's Photomerge script, creating a mashup of deities, talismans, stellae, gods, scribes and statues. The execution was fast, and resulting images were printed and covered an entire wall for the exhibit. The images had a photorealistic quality, yet presented objects that could never exist; in part because human history has never created such icons, in part because, like an M.C. Escher drawing, the colors, space and structures depicted don't quite add up.

Iconoclashes is unusual in that it had a subsequent life in multiple gallery shows after the event. Apparently, what is created at a hackathon can have an enduring legacy! It also demonstrates that fully 'finished' artwork can get created even under significant time constraints.

A subsequent theme, 'Deluge' (NYC 2016)[15] explored the role (or non-role) of artists and technologists in fighting climate change. The event took place in New York's Red Hook district, an area that was flooded during Hurricane Sandy. The deluge is of course not merely literal: we are awash in technology that threatens to subsume us.

Glacier Pop

It's not uncommon for participants to try something too ambitious, only to realize at the last moment that their original idea won't work, or that they won't have time to complete it. During 'Deluge,' this happened to artists and

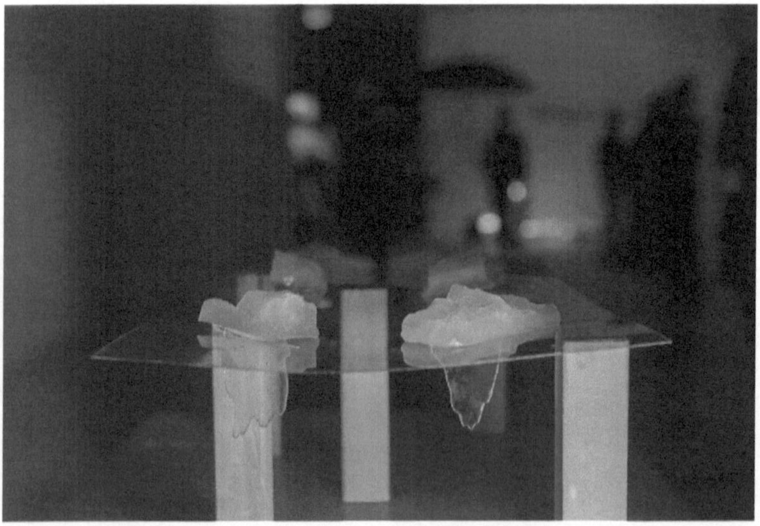

FIGURE 14.3 *Glacier Popfrom* from Art Hack Day Deluge. NYC (2016). Credit: Margarida Malarkey, Art Hack Day.

technologists Anne-Marie Lavigne, Igal Nassima and Sue Ngo. Fortunately, they had a backup plan, one that was not only analog but also edible.

They made popsicles in the shape of floating icebergs, entitled 'Glacier pops.'[16] Apart from being a tongue-in-cheek reference to the appetizers served during upscale gallery openings, the work was also a metaphor for our voracious appetite and resulting damage to the environment. Visitors to the opening were invited to eat the popsicles, and bring the popsicle sticks back home, also iceberg-shaped, as a reminder of the issues we are facing regarding climate change. Any iceberg popsicles that weren't eaten, melted away during the exhibit, mirroring the state of the ice caps.

Glacier Pops is an example of the type of ephemeral work that the one-night exhibit and format encourages, work that might otherwise not be created.

Conclusion

Art Hack Day strives to catalyze new collaborations by providing a playful, resourced and structured format, which resulted in artworks that were critical, playful, technically accomplished and daringly analogue. Whereas contemporary art shows can feel self-contained and self-sufficient, and serve a particular kind of audience, *Art Hack Day* tries to democratize both art and hacking for a broader and more diverse community. This is achieved via the perceived spontaneity of the format, fostering a productive atmosphere of unbridled experimentation and collaboration and mindfully engaging with people of different backgrounds and practices to participate in *Art Hack Day*.

Art Hack Day seeks to explore how technology can facilitate expressiveness and new ways of being in the world, but also the power of radical collaboration in art. In the context of *Art Hack Day*, authorship is a diffuse, perhaps even unnecessary, notion. We have found that attributing authorship is more often than not an exercise in futility. We focus on the process and context in which art emerges, not on the persona of the artist (although we recognize that there are important and legitimate economic reasons to do so).

The hackathon format, although adapted to the art world, doesn't carry over seamlessly. While *Art Hack Day*, with its open source ethos, aims to challenge capitalism as a dominant paradigm, it can also be seen as undermining the very livelihood on which professional artists depend, in particular when the event is associated with a major, funded, institution or festival.

Art Hack Day is ephemeral in nature. Whether or not the works created during *Art Hack Day* have a lasting impact is an open question, and perhaps it doesn't matter. Perhaps the most enduring legacy of *Art Hack Day* won't be the artworks created, but the human relationships and many lasting personal friendships that so far have sprung from the event, strengthening the nascent practices at the intersection of art and technology.

Notes

1 A disclaimer: There are a number of topics, however pertinent they might seem to hacking, that I won't discuss: 'The Hacker ethic', as articulated by Steven Levy, the impact of the 'Hacker manifesto', the MIT Railroad Club of the 1950s (often seen as the place where 'hacking' was born), Eric Raymond's seminal book on open source culture: "The Cathedral and the Bazaar", Paul Graham's book "Hackers and Painters", nor will I discuss the recent phenomenon of artists rebranding themselves as hackers (partially due to the cultural cachet the term acquired, and impact the likes of Edward Snowden and Julian Assange have had on mainstream culture). There's also the opposite, if less visible, phenomenon, whereby successful technologists go into art, after having made a fortune at the latest breakout startup.
2 http://musichackday.org
3 The Consumer Electronics Show: https://en.wikipedia.org/wiki/Consumer_Electronics_Show
4 https://en.wikipedia.org/wiki/Application_programming_interface
5 Manufactured by Anybots, http://www.anybots.com/
6 https://www.sxsw.com/
7 https://arambartholl.com/speed-show/
8 http://www.byobworldwide.com/
9 http://fffff.at/
10 http://grayarea.org/
11 http://arthackday.net
12 https://transmediale.de/archive/history/festival/2014
13 http://www.arthackday.net/events/god-mode
14 http://arthackday.net/projects/erik-berglin-clement-valla-iconoclashes
15 http://arthackday.net/events/deluge
16 http://arthackday.net/projects/susan-ngo-igal-nassima-anne-marie-lavigne-glacier-pop

+ CONTEXT V

Museums, Galleries, Festivals and Programs

15

CRITICAL MAKING AS A MODEL FOR CURATING OR MAKING EXHIBITIONS AS THINGS TO THINK WITH

Nora O. Murchú

Introduction

Our world has become increasingly complex. How we perceive and understand it is continuously subject to processes of globalisation, the expansion of capitalism and scientific and technological developments. The interrelatedness and unpredictability of these complex systems – from computer networks to society and social networks, and from financial markets to the climate system – makes it difficult to understand the world from any one dominant structure or model of thinking. Instead, this growing complexity and how its connections and relations constitute our understanding of things, points to an approach where the intersection of science, art and philosophy 'sees all things together'[1] to provide means to formulate questions and understand the world.

As technology plays an increasing role in our lives, artists are undertaking a process of opening up and extending our critical understanding of it. Engaging in processes of making and tinkering, artists are experimenting with material arrangements of data, algorithms, hardware and software to articulate particular stances and ideas through their production, and by making transparent the processes and thoughts that underpin their construction. These artworks-in-progress are often disseminated through Twitter feeds and Facebook posts, where the public display of these works mediates exchanges between artists, their peers and audiences. These exchanges allow for the reflection on the personal and social impact of these technologies, often evoking further ideas of inquiry for artists to produce further adjustments to the artwork. This process of making, sharing and learning highlights how artistic practice can be increasingly seen as a process that engages in prototyping or tangible instantiations of evolving ideas and inquiry.[2]

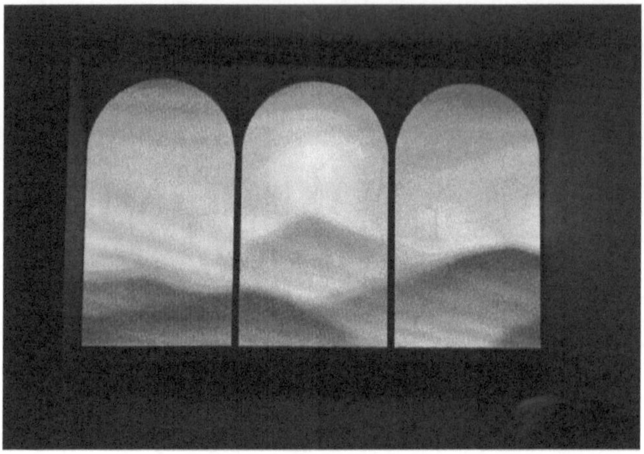

FIGURE 15.1 *RGB Landscape*, Nicolas Sassoon in Espace Multimedia Ganter, France. All photos in this chapter by Nora O Murchú.

This shift in the art object – from an object with a fixed arrangement of meanings, material and aesthetic composition to one that is open and subject to continuous flux – points to an emerging model of practice where a culture of prototyping and making is becoming central to how artists are producing their work. Making has come to the foreground of academic, business and artistic research in the past few years, and studies of maker identities, behaviours and their communities have contributed to the understanding of the cultural impact of the practice. Through appropriation and the manipulation of technological materials, the maker engages in a 'process of building, designing, and innovating with tools and materials'[3] to alter everyday situations, objects and rules to build provocations that encourage a re-evaluation of technology in culture. Over the past few years, artists have developed a broad repertoire of subversive strategies and aesthetic forms affecting all art forms, genres and age groups. Considering this process as a framework for making art allows for a number of possibilities: firstly, it allows artists to unfold their ideas over time; secondly, it invites reflection on the material arrangements of the artwork produced throughout this process and finally, it allows for this process of prototyping to be seen as a strategy for engagement that opens up possibilities for both artistic and curatorial actions.

The curator once functioned solely within the museum and was entrusted with the overseeing of a particular collection or display. However, as artists engaged with technological tools and the internet as part of their practice, curators too began engaging with these modes of production for curating. In the past few years, curating has undergone significant changes – working inside and outside of institutions, online in networks and distributed systems – curatorial work has continuously evolved to support artistic work that challenges the traditional understanding of interaction, participation and collaboration within wider cultural systems.[4] But what modes of display and exhibition can support artistic practices that engage with making and how might curating learn from them? As outlined above, artists who engage in making have a new set of conditions by which they make their work – through which the political qualities of an issue are materialised by participatory means.[5] This is an emerging context for curatorial work and opens the possibility for a culture of critical, informed and reflective practice that incorporates new strategies and critical approaches from making.

Like making, curatorial work occurs in many overlapping contexts: technical, artistic and institutional – each of these contexts introduces new concerns and constraints on possible methods and audience engagement. For example, curators are increasingly experimenting with online platforms to create exhibitions that account for the network characteristics of internet-based work, allowing audiences to experience this artwork at the browser interface. These exhibitions are significant for building knowledge about curatorial practice that engages with digital technologies. It shows how curating can take place in new contexts by linking the building of exhibitions to both the technical and material conditions of artworks. It also brings an audience awareness of technological practices and demonstrates how curation can extend knowledge in relevant technical areas by providing the means for the conceptual exploration of them.

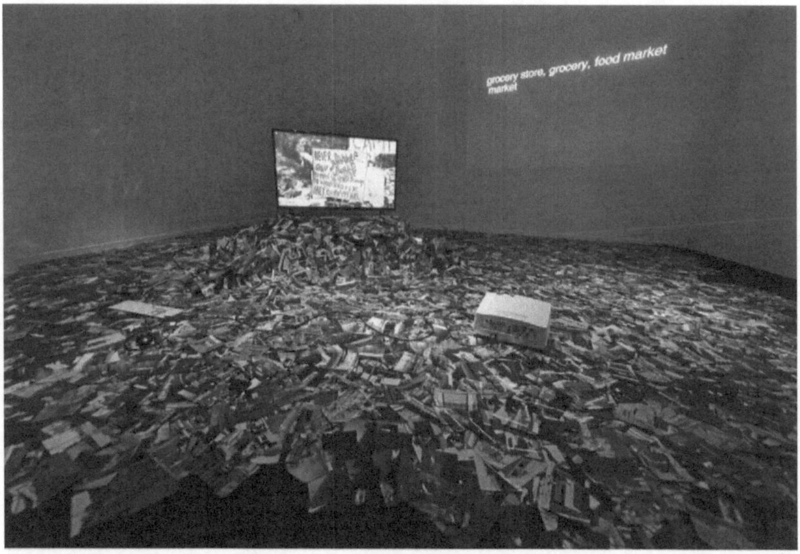

FIGURE 15.2 *You Can Look and Understand it, Without Knowing What It Actually Is,* Alan Butler in France.

As a grounded practice with material and conceptual engagement, making focuses on the exploration of values in society and their implementation within technology. Similar to curating, it is a way of thinking through what (and how) technologies mean by understanding how they work, and their related academic concepts. The maker engages with the material layer of digital technologies through prototyping to interpret and to intervene in the values embedded within them. The outcomes of this material thinking are intended as socio-technical critique, created to test or materialise hypotheses in public.[6] 'The moment these objects are placed in an art context, it's clear they are to be engaged with critically, instead of deployed,'[7] allowing for the rethinking of designed objects and new systems of public knowledge to emerge. This approach to art-making can offer new potential for curatorial intervention, not limited to encouraging the audience in developing subjective interpretation but also introduces a set of alternative ways of considering the artwork, beyond how an audience traditionally encounters an exhibition. If an artist prototypes objects, to what extent does curating prototype an exhibition and connect it to processes of shared construction, joint conversation and reflection?

Prototyping Exhibitions

Curatorial work is characterised by multiple forms of production and dissemination: collecting, conserving, displaying, contextualising, critique and publicising. Exhibition-making has long been considered one of the key aspects

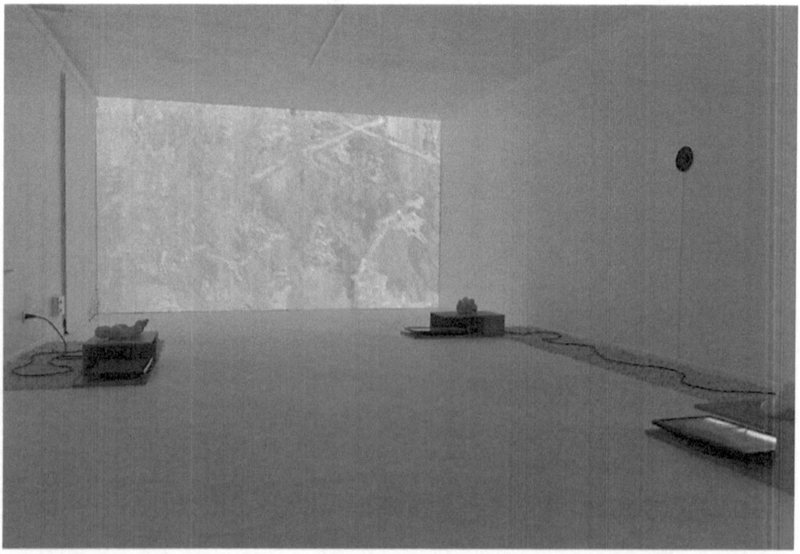

FIGURE 15.3 *Neural Network Landscape*, Gregory Chatonsky in France.

of curating and has undergone significant reinvention over the past few years. Exhibitions operate not only as spaces of presentation and display but also as a means to formulate ideas about artistic production and to transform its reception.

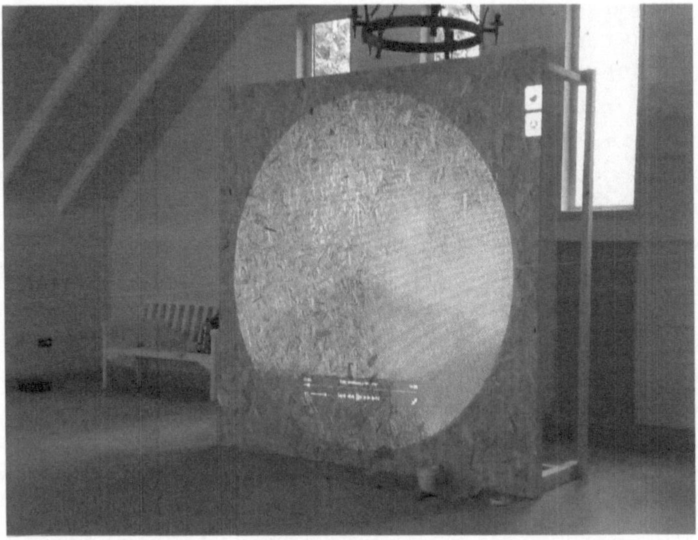

FIGURE 15.4 *RGB Landscape*, Nicolas Sassoon in Latvia.

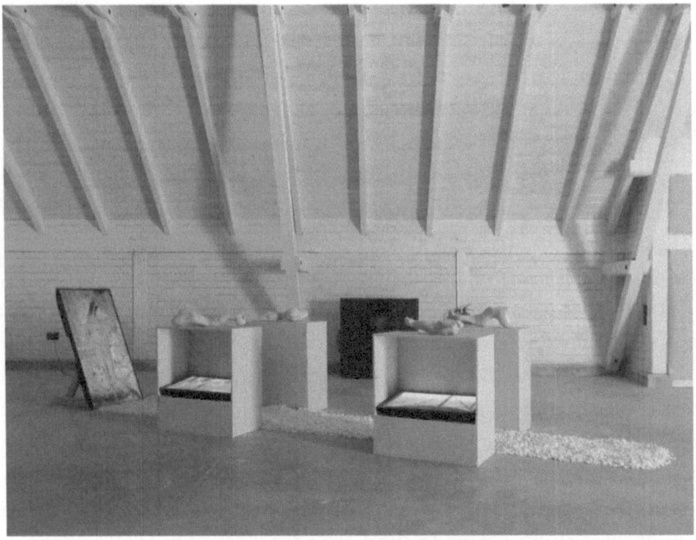

FIGURE 15.5 *Ossuaires*, Gregory Chatonsky in Latvia.

Through the exhibition, it is possible for works to establish relationships not only with an audience but also with other works and the discourse of the historical trajectory of art. The exhibition operates not only to provide discourse or context but also to represent knowledge-in-action and act as a means for public understanding and learning. But how do curators produce meaning in the physical process of making an exhibition? As a form of artistic interpretation, the curatorial work of creating an exhibition is traditionally seen as grounded in the historical conventions of exhibition display or representative of a curator's understanding of an artwork. However, this reduction of curatorial work to an exercise of arranging artworks based on historical conventions of display overlooks how artworks are experienced by their audience, and the meaning-making a curator engages in through processes of inquiry and experimentation within the gallery.

In planning and installing an exhibition, curators frame artworks and build meaning based on contextual resources they have available to them. In exhibition installation, this includes the artworks, the gallery space, lighting systems, technical equipment and the time to move objects around. For curators to achieve what 'feels right' in the making of an exhibition, they will place and move artworks within the gallery based on conceptual, aesthetic and discursive features of the artworks, while reflecting and experimenting to explore how meaning can be constructed and understood. For example, when selecting artwork for the exhibition *Land / Sea / Signal*, I initially had selected two artworks that upon closer examination were visually similar in terms of the colours used within them. My objective for the exhibition was not to draw comparisons between the artworks based on their aesthetic features. Instead, I aimed to create connections between the discursive themes of the artworks. This led me to select a different artwork from one of the artists to ensure that audiences would not assume that one artist had authored both artworks. Thus, the physical making of an exhibition can be seen as a combination of 'plans and situated actions'[8] where the curator engages in a process of inquiry and meaning-making through experimentation with modes of display and artwork arrangement within the gallery. Consequently, an exhibition is the outcome of curatorial activities of inquiry and action and grounded in the comprehension and nature of curatorial practice. From this perspective, it is possible to consider an exhibition as a prototype in that it can be seen as a version of a curatorial idea – something flexible and changeable – a means for learning something about the effects of the conceptual understanding, material arrangement and technical presentation of artworks on an audience. The outcome of this experimentation often has a consequence for the public understanding of the exhibition, particularly when the result of the installation presents artistic knowledge that differs from that in exhibition catalogues or accompanying curatorial essays. This demonstrates how exhibition making can be framed as a prototyping process that encourages the exploration, analysis and mediation of artworks and their artistic concepts with the aim to develop a new understanding of them.

Land / Sea / Signal

Commissioned as part of a Creative Europe Cooperation project between Rua Red South Dublin Arts Centre with Le Département du Territoire de Belfort in France, and Rezeknes Novada Pasvaldiba, in Latvia, *Land / Sea / Signal* directly addressed the complex socio-political conditions embedded within internet infrastructure. The exhibition toured between the three venues and included artists (John Gerrard, Alan Butler, Gregory Chatonsky, Nicolas Sassoon and Santa France) from each of the respective countries of the project partners (a stipulation from the initial proposal, which aimed at building networks between the three partners). The exhibition aimed to make evident how humans, technologies and our environment intersect, and initially, set out to call attention to the invisible ways in which technology intersects with the everyday through infrastructure and software. When considering artists for inclusion, I selected those whose work reflected on these concerns, and through their practice exposed the inner-workings and external influences to these systems, and their increasing authority in society. Hence, it was necessary to include artists whose work critically examined the present conditions of these systems and experimented with material arrangements that attempt to perceive them.

In my role as curator, my objective was to draw together both procedural and cultural interpretations of these systems. To do this, I considered how the theme of the exhibition, the artwork and the physical gallery space formed together to support the interpretation of the exhibition and act as a discursive space for the audience. Each of the resulting exhibitions was the result of a process of constant negotiation between thinking, making and reflecting. By experimenting with these aspects of the curatorial process, three versions or prototypes of the exhibition emerged with distinct values, meanings and material forms from one another. The exhibitions emerged over a year-long research endeavour and were refined and developed over the touring period. During this period, each of the exhibitions were subject to an exploratory process of perceiving, reflecting, structuring and problem solving, from which I was able to generate new insights to the artwork through the continuous refinement of the material ways in which the artworks were contextualised in space. Each instantiation of the exhibition as it toured offered an opportunity for refinement and understanding of this theme, allowing me to familiarise myself further with the material dimensions of the artists' work, and reflect on the final arrangement of the exhibition to assess the impact on the audience. It also offered iterative and continuous consideration of the multiple ways in which infrastructure and their inter-acting relations perform.

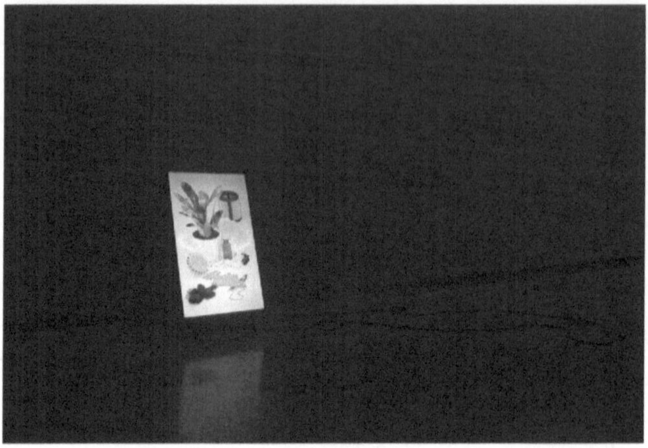

FIGURE 15.6 *A Matter of Respiration*, Santa France in Dublin.

The first of these exhibitions took place in Espace Multimedia Ganter in France, where three of the artists were present for the installation process. Initially, a plan in consultation with the artists was determined in advance to the installation process, from which minor alterations were made in-situ. During the installation process, the artists and curator reflected on the physicality of the gallery space, and discussed the relations between the artworks including the modes of display that were specific to each of the artworks. As a result, I gained insights into the ways in which artists tied the production of their artworks directly to their modes of display. For example, in the work of Santa France, the artist developed three videos using 3D modelling software. Within the software she used to develop the videos, the interface is split in four, and each pane offers a different perspective of the objects being created. Consequently, the artist created the videos based on these perspectives, and each of the videos was displayed on a TV oriented in a particular way – on the floor, leaning against a wall and on one of the gallery walls. Through interacting with the artists during the installation process, my knowledge and understanding of how each of the artworks should be encountered was developed further.

In contrast with the exhibition in France, the second exhibition in Latvia was significantly challenging due to the physicality of the space – a newly constructed barn that had beams extending into the space. As a result, the artwork by Nicolas Sassoon needed to be installed differently for the exhibition. *RGB landscape* was originally projected onto a wall in France. However, due to the architecture of the space it was not possible to display the work in the same manner. In conjunction with the artist, I discussed their preferences for the display of the work (perceived and communicated through their emergent relationship to the film) with what worked best in the space, as well as with the artist's concept for the piece. This resulted in the construction of a wooden structure for projection. The structure was made to the artist's specification. However, having not seen the architecture of the gallery prior to the installation, the scale of the structure posed problems. During the process of achieving a satisfactory 'look and feel' for the exhibition, I experimented significantly with the arrangements of artworks within the space, moving them around until I was satisfied with 'what worked' for the exhibition and how an audience would interact and view the work.

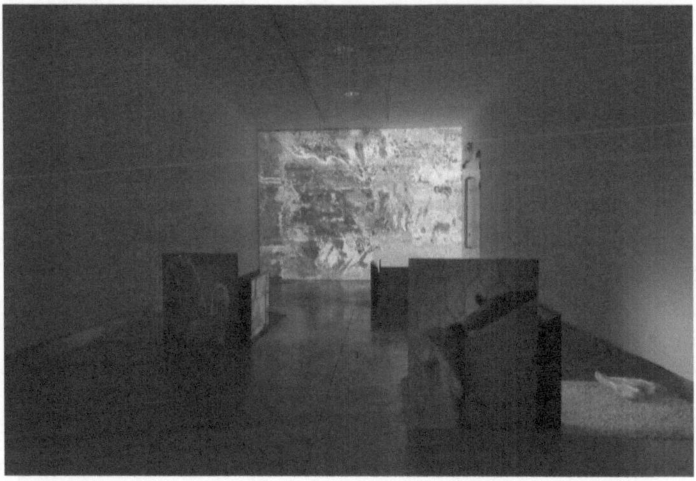

FIGURE 15.7 *Neural Network Landscape*, Gregory Chatonsky in Dublin.

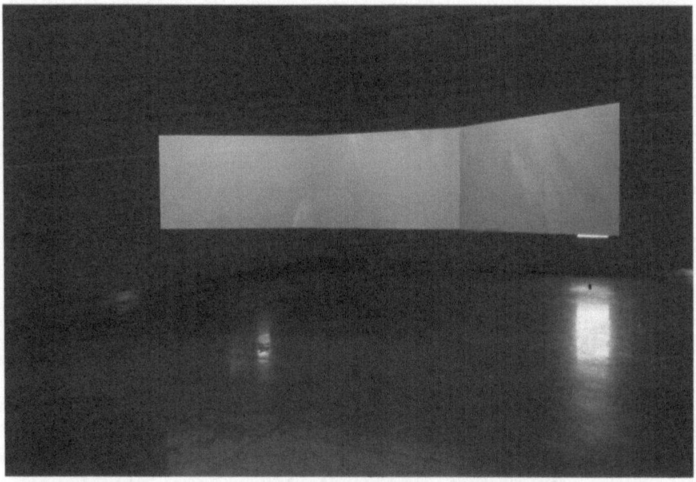

FIGURE 15.8 *Signals*, Nicolas Sassoon & Rick Silva in Dublin.

The Dublin iteration of the exhibition aimed to embody the exhibition narrative and consider how an audience could enter into meaning-making activities through physical orientations to the artworks. Having gained an in-depth understanding of the artworks, and reflected upon the issues in the previous exhibitions, each artwork's mode of display was reconsidered. Based on conversations with the artists, a number of material changes were made and the exhibition was designed to lead the audience in a particular direction.

Conclusion

This chapter points to the continuing urgency of the curatorial field, its growing complexity and new ways of thinking that encapsulate things coming together. The aim of thinking of curating as a form of critical making is to understand how continuous recombinations of people, data and machines might offer new understanding and insight into how technological concepts shape our everyday – particularly when engaging with research and exploring emerging technologies. This approach can disclose new potential and leave wider room for curatorial intervention, not limited to encouraging the audience in developing subjective interpretation, but deliberately introducing a set of alternative ways of considering and acting within the work, beyond the traditional codes and standards typical of an exhibition context.

Notes

1 Von Foerster, Heinz. n.d. "Constructions of the Mind." *Stanford Humanities Review* 4 (2). https://web.stanford.edu/group/SHR/4-2/text/interviewvonf.html [Accessed January 26, 2019].

2 Arrigoni, Gabriella, and Tom Schofield. 2015. "Understanding Artistic Prototypes between Activism and Research." 24–39. Glasgow, Scotland.

3 Litts, Breanne K. 2015. "Making Learning: Makerspaces as Learning Environments." Ph.D. Thesis, University of Wisconsin-Madison.

4 Graham, Beryl, Sarah Cook, and Steve Dietz. 2010. *Rethinking Curating: Art after New Media.* Leonardo. Cambridge, MA; London, England: MIT Press.

5 DiSalvo, Carl. 2014. "Critical Making as Materializing the Politics of Design." *The Information Society* 30 (2): 96–105. doi:10.1080/01972243.2014.875770 [Accessed January 28, 2019].

6 Arrigoni, Gabriella, and Brigitta Zics. 2016. "Fiction and Curatorial Practice: Developing Alternative Experiences for Digital Artistic Prototypes." *International Journal of Performance Arts and Digital Media* 12 (1): 82–94. doi:10.1080/14794713.2015.1133960 [Accessed February 1, 2019].

7 Hart, Sam. 2014. "Critical Engineering: An Interview with Julian Oliver Avant. Org." http://avant.org/artifact/julian-oliver/ [Accessed February 1, 2019].

8 Suchman, Lucy A. 1987. *Plans and Situated Actions: The Problem of Human-Machine Communication.* 2nd edition. Cambridge; New York: Cambridge University Press.

Bibliography

Ferguson, Bruce W., Reesa Greenberg, and Sandy Nairne, eds. 1996. *Thinking About Exhibitions*. 1st edition. London, England; New York: Routledge.

Ratto, Matt. 2011. *Open Design and Critical Making. Open Design Now: Why Design Cannot Remain Exclusive*. Amsterdam: BIS Publishers. http://opendesignnow.org/index.php/article/critical-making-matt-ratto/ [Accessed January 26, 2019].

Ratto, Matt. 2009 "Critical Making: Conceptual and Material Studies in Technology and Social Life." Hybrid Design Practice workshop, Ubicomp, Orlando, FL

16

WIKITOPIA HONG KONG

Curating a Collaborative Urban Future

Yidi Tsao

WIKITOPIA is a media arts and culture festival in Hong Kong that celebrates creative interventions mediated with digital technologies as collaborative contributions to the future of society. This chapter examines the three iterations of WIKITOPIA produced to date (2010, 2012–2013 and 2016). These serve as case studies to investigate how the open wiki platform was translated offline as a kind of hacking of curatorial practice. Throughout, I reflect upon the origin and execution of the event within the particular social, cultural, political and economic context of Hong Kong City. To begin, I will sketch out the conception of the festival. This discussion is anchored in an interview that I conducted in 2018 with Hong Kong-based curator and artist Ellen Pau. Inspired by the rise of hacker culture and the maker movement in Hong Kong in the first decade of the 21st century, Ellen began to envisage the festival as early as 2009 when she returned to Videotage after a two-year hiatus. A discussion of the festival's origins is followed by programming strategies and technical difficulties that curators encountered in planning each of the three editions. I personally worked on the third iteration of WIKITOPIA in 2016, which focused on the topic of surveillance and privacy and was in collaboration with Goethe-Institut Hongkong under the framework of the International Symposium on Electronic Arts Conference (ISEA 2016), Cultural R>evolution 香港 Hong Kong.

Conception: Questioning Emerging Paradigms

WIKITOPIA is the brainchild of Ellen Pau, who is considered to be the 'godmother' "教母"[1] in Chinese) of the media art scene in Hong Kong, due to the tremendous effort she put into building the scene and mentoring the younger generation. In addition to her day job as a hospital radiographer, Ellen has been active in the Hong Kong art scene since the 1980s, as a self-taught video artist, cinematographer, curator and art critic. She co-founded Videotage in 1986 and is Founding Director of Microwave New Media Art Festival (began in 1996). Videotage and Microwave are the trailblazers in the region that evolved from presenting at first only video art, then a wider spectrum of new media artworks and dialogues on digital culture and society.

'Wiki' in the title of the festival 'WIKITOPIA' derives from the 'WikiWikiWeb server' concept created by programmer Ward Cunningham in 1994. Today we are familiar with the term Wiki thanks to Wikipedia, the free and accessible online encyclopaedia that allows users to both read and edit content. Lesser known than the online encyclopaedia, but equally important, Wiki technology is implemented on many websites that require collaboration, such as the SCP (Secure, Contain, Protect) Foundation.[2]

As Leuf and Cunningham writes,

> A wiki is not a carefully crafted site for casual visitors. Instead, it seeks to involve the visitor in an ongoing process of creation and collaboration that constantly changes the Web site landscape.
>
> Wiki is a lot about a *collaboration space*, albeit an unusual one because of its total freedom, ease of access and use, simple and uniform navigational conventions, and apparent lack of formal structure. Wiki is also a way to organize and cross-link knowledge, perhaps its main purpose for the single-user wiki.
>
> Wiki is *inherently democratic* — every user has exactly the same capabilities as any other user.[3]

When Ellen Pau conceived of the first edition of WIKITOPIA in 2009, this year also marked the 12th anniversary of the transfer of sovereignty over Hong Kong from the United Kingdom to the People's Republic of China. Under British colonial administration, cultural policy in Hong Kong was short-sighted, while 'a transition to an attempt of centralised control' had been in place since the handover.[4] Though Hong Kong was promised to remain a democratic society for 50 years after the handover, people were, and continue to be, worried about the possibility of a totalitarian government taking over in the future. This anxiety is felt even more by artists and cultural practitioners since without freedom of speech or expression, the art and culture scene will no longer be the same due to censorship of certain topics[5] and even worse, self-censorship.[6]

Note that in addition to the change in political atmosphere, there were also new players in Hong Kong around the time when WIKITOPIA was conceived. Creative Commons Hong Kong had just been founded 'to promote the use of Creative Commons licenses in Hong Kong.' Their stated goal is 'to serve Hong Kong's diverse creative talents' as 'nourishing free culture and legal sharing can greatly enhance our educational environment, as well as boost Hong Kong's capacity for innovation and creativity.'[7] Another was the start of The Good Lab, Hong Kong's first Social Innovation hub, which aims to 'build a community of changemakers' and to 'ignite interest and encourage collaboration between sectors to solve complex social and environmental problems.' All of these new initiatives contributed to a rising recognition by Ellen Pau and others for the potential of implementing the 'wiki' concept in the offline world as a democratic medium and an engine for social change. Early thoughts about what would become WIKITOPIA raised questions such as: what can an individual do? Where can a citizen begin confronting some of these underlying concerns? Collaborators began considering how they could create a festival to celebrate emerging ideas in digital culture that would address new issues openly.

> The appeal of wiki technology lies in the act of rethinking the familiar. Once again, hierarchical control models are at our disposition, and with them, valid ideas of why and how, through the division of labor, complex problems can be solved and products produced and distributed.[8]

Wiki suggests not only collaborative modes of design, revision, implementation and reiteration on the Internet but also challenges conventional concepts of ownership, objectivity and hierarchical forms of work and social structures. It brings a certain level of 'culture shock' to its first-time users, as the working and organisation methods are fundamentally different from our current top-down approach in most institutions and in day-to-day life.[9]

Today, Shenzhen, Hong Kong's neighbouring city across the border[10] in China, is heralded as a 'Mecca for makers' with cheap electronic parts and an unbelievably fast prototyping capacity. Just a decade ago, Shenzhen was infamous for its 'Shanzhai' products.[11] This phenomenon has been dismissed and criticised by many as mere copycatting, while others, such as *The Wall Street Journal*,[12] embraced it as 'the sincerest form of rebellion,' a type of 'hacking' when considered in a broader context. As stated by Luisa E. Mengoni, Head of the Victoria and Albert Gallery in Shenzhen, 'the free, informal and open source systems through which both makers and Shanzhai manufacturers operate has allowed the development of a particular type of grassroots innovation, particularly in Shenzhen.'[13] The Victoria and Albert Museum, in collaboration with the Shenzhen local government, conducted field research into local making communities and the manufacturing scene before their gallery in the new Design Society museum and culture hub in Shenzhen opened its doors on 2 December 2017.

Structures of power have always existed in the art world that artists must work within or fight against (or vacillate between both stances). At the most extreme, we still see a 'Power 100' list[14] produced by *ArtReview* annually, selecting what it views as the 'most influential people in the contemporary art world.' Much of the mainstream art world is tightly coupled with capital and social status rather than creativity and society. To many artists, this contemporary notion of art has become too institutionalised to fulfil their desire for creative output. In considering these systems, the curators imagining WIKITOPIA asked: as a field that produces (or used to produce) the most radical ideas, can art embrace a democratic and decentralised world without relying on hierarchy? Can this be encouraged by applying wiki methods within curatorial practice?

In 2008, ARTHK[15] was newly launched by cultural entrepreneur Magnus Renfrew, who is also about to launch a new art fair in Taipei.[16] ARTHK soon drew the attention of international galleries and was later purchased by Art Basel Group as a strategic step of Art Basel's global expansion. This was followed by international galleries opening franchises in Hong Kong that don't necessarily benefit the local art community in sustainable ways. In this climate, the idea of 'WIKITOPIA' was born in 2009. It would be an occasion where people from many walks of life ranging from bankers to professors to hackers, who are affected by digital culture, who are knowledgeable and interested in social change, could talk, inspire, learn from and provoke one another.

Content and Format

Dubbed 'the first TinyFest on collaborative future in Hong Kong,' the first edition of WIKITOPIA was launched in September 2010, spanning two weekends with different types of events including screenings, keynotes, panel discussions, workshops and parties. To kick-start the festival, a two-day workshop *Urban Invader* was designed and hosted by Hong Kong new media artist Keith Lam, in which he preached DIWO ('Do It With Others', a term coined by Furtherfield in 2006[17]) culture by equipping participants with technical know-how in electronics and by encouraging them to reflect upon relationships between urban development and light pollution or air quality. The outcome was a practical gadget that participants could use to visualise information on the street and to raise others' awareness to problems incurred in rapid profit-driven changes to the cityscape. Also included was 'Why-Fi?' by Dutch hacker moddr_ who demonstrated the fragility in wireless network security by teaching people how to crack Wi-Fi encryptions.

These workshops fulfil the most important purpose of realising WIKITOPIA as a festival, a medium that is less often adopted in Asia than in Europe. By directly engaging audiences, the workshops encouraged active participation and interaction amongst every attendee. Keynotes and panel discussions, though informative and insightful, are more of a one-way, didactic method that is not intrinsically 'wiki.' The workshop format raises a question to curators: why do we still need exhibitions? For example, considering Keith Lam's artwork, his message comes across more clearly to audiences when they are active

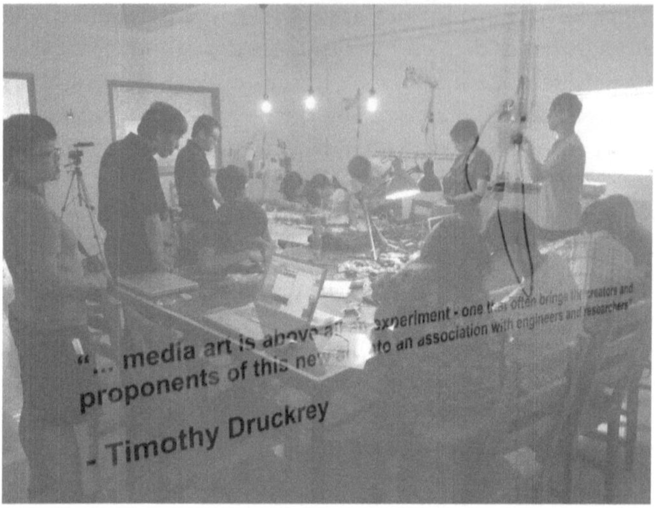

FIGURE 16.1 *Urban Invader* workshop by Hong Kong artist Keith Lam, 11–12 September 2010. Photography by Wong Sin Man Mandy.

participants rather than passive onlookers. When the goal of an artwork is to raise awareness in citizen-viewers and when we can demystify the making of an artwork by demonstrating hands-on practice, wouldn't workshops be a more suitable framework for exhibiting media artworks?

In addition to hands-on workshops, WIKITOPIA 2010 also highlighted 'Wiki,' or crowdsourcing, as a method of artistic creation. Brazilian artist Kika Nicolela presented the *Exquisite Corpse Video Project (ECVP)*. ECVP is a collaborative networked video art creation that involves artists from all over the world. Inspired by the surrealists' 'Exquisite Corpse' game, ECVP artists embed the last ten seconds of a video from a previous member and create their own video in response. Other crowdsourced video artworks screened during the festival included Perry Bard's *The Global Remake*, remaking the 1929 experimental silent film *Man with a Movie Camera* with reinterpreted footage of any of the 1,276 shots in Vertov's classic, streamed on the site alongside the original. There was also Sergei Eisenstein's *Battleship Potemkin*, which was reproduced by 105 students from Yildiz Technical University as project *re_potemkin*.

Electronic musician and DJ, Pogo, is known for using Internet video sampling to produce music videos that he shares on YouTube. Together with local music collective iii, Pogo was featured at a party where participants could appreciate the music created in a more relaxed way. These performances and the event context brought forth discussions on collaborative modes of artistic creation, intellectual property, copyright, the copyleft movement, open source projects and online crowdsourcing.

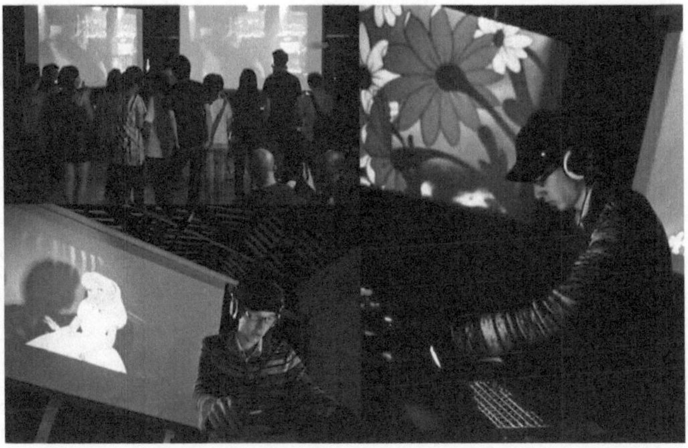

FIGURE 16.2 DJ Pogo at Audiovisual Mashup Party, WIKITOPIA, 19 Sep. 2010. Photography by Wong Sin Man Mandy.

The post-event catalogue of the second edition of WIKITOPIA states that,

> The festival aims to act as the pivotal point for artists, ICT personnel, media activists, curators, writers, thinkers and any curious minds to exchange insight and knowledge through a diverse range of programmes in town. This year, WIKITOPIA's theme Data Gaga focuses on data as art, network technology, and the collaborative creativity of the Maker Culture.

Two years after the first edition, as smart tracking devices prevailed, data had taken on an increasingly central role in everyday life. Since 'data becomes the new oil,'[18] we aimed to find ways that it wouldn't create only data tycoons but it would empower everyone.

In November 2013, the ISEA International committee announced that Hong Kong won the bid to host ISEA 2016. Half a year before that announcement, at Mira Hotel in Hong Kong, Edward Snowden revealed surveillance schemes run by the US government. This incident drew my attention, along with the world's, to topics of surveillance and privacy. My own interest, however, is partially due to a cultural difference between China and the United States. In China, mass surveillance has long been viewed as widespread government practice; everyone is aware of it, but no one calls it out as Snowden did. In April 2015, *Citizenfour*, the documentary about Snowden by Laura Poitras,[19] was screened in Hong Kong at Broadway Cinematheque in Yau Ma Tei. At the Q&A session, I shared my doubts about the film. For cultural and historical reasons, especially when viewed in light of Confucian and communist values, people in Hong Kong and China understand the concept of privacy and the purpose of government differently than those in the West. This may be why the topic is rarely publicly discussed in Hong Kong and in Mainland China. Because of this discrepancy, I thought it would be useful to bring people together to voice their opinions on the issue. Our vision resonates with the tremendous controversy caused by Baidu CEO Robin Li's recent comments on how the Chinese are willing to trade privacy for convenience.[20]

When examining national social policies, it is important to bring in perspectives from other jurisdictions and cultures. In planning WIKITOPIA 2016 in conjunction with ISEA, I proposed the topic to Goethe-Institut Hongkong, inviting them to collaborate on the festival. I chose to approach the Goethe-Insitut was because it was apparent to me that people in Germany are on the opposite end of the spectrum when it comes to issues of privacy – consider the public opposition to Google Street View.[21] The result of the Goethe-Institut collaboration included a keynote on 'Digital Self-Defence, the Chaos Computer Club between Hacking, Activism and Courts' delivered by Linus Neumann, the spokesperson for Chaos Computer Club (CCC). The CCC is 'Europe's largest association of hackers… providing information about technical and societal issues, such as surveillance, privacy, freedom of information, hacktivism, data security and many other interesting things around technology and hacking issues.'[22] Other events included the screening of *The Lives of Others*, a tour around Hong Kong Polytechnic University campus where the festival took place, mapping out surveillance cameras and sending a request to access the footage to the university's security team; a workshop to become 'Digitally Invisible' with artist/scientist Lasse Scherffig; and the 'Me and My Shadow' workshop with Tactical Technology Collective to get one's digital shadows back under control. Finally, CryptoParty, hosted by Hong Kong hackerspace Dim Sum Labs, was an event to pass on knowledge about protecting oneself in the digital space.

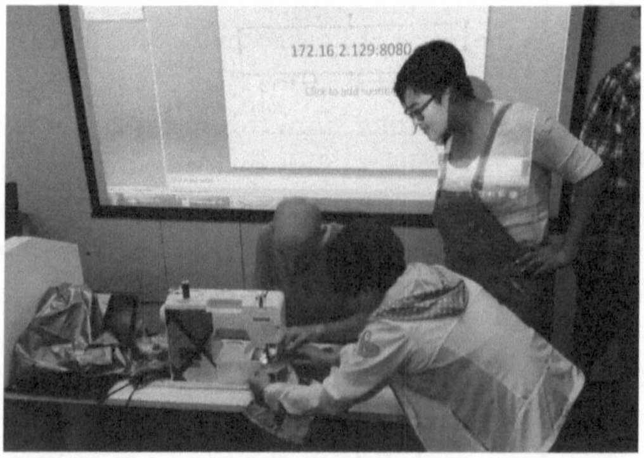

FIGURE 16.3 Workshop by artist/scientist Lasse Scherffig. Photography by Christopher LEE.

These events and workshops underscored the themes of WIKITOPIA 2016, celebrating digital privacy and interrogating the quick spread of mass surveillance schemes in the name of security, as fuelled by creative interventions. One project that we presented by a Hong Kong-based practitioner was Access My Info (AMI),[23] a Web application developed by Dr. Lokman Tsui from Chinese University of Hong Kong with In-media, Keyboard Frontline, Open Effect and the Citizen Lab at the Munk School of Global Affairs (University of Toronto). With this app, users can 'request access to their personal information with their telecommunication companies or Internet service providers (ISPs).' AMI sheds light upon the grey zone of personal information protection that users tend to forget when enjoying the convenience brought by the use of digital devices.

Broadcast and Documentation

The ever-increasing speed of the evolution of media technologies becomes central to the creation and curation of media arts. The first two editions of WIKITOPIA were recorded and live-broadcast online to erase any geographic limitations of the physical event. In 2010, however, live streaming was not as easy as it is now. For example, the current 'Facebook Live' function (which was launched in August 2015 but initially only for celebrities[24]) is now free and accessible to all (with an agreement to Facebook's terms of service – the content must of course obey 'community standards'[25]). In 2010, however, the WIKITOPIA team had to commission a company specialised in live streaming events to make this aspect possible. The cost of this increased the festival budget significantly. We do not have the statistics of how many people actually watched the live stream,[26] so it is not impossible to evaluate its impact. We felt that the role and presence of the livestream complemented the festival concept and that certainly, with the convenience and low cost of the technology today, live streaming should remain the norm in future editions of WIKITOPIA.

By WIKITOPIA[27] 2016, the technical aspect of live streaming became less of a problem while an ethical discussion around recording was central to the philosophical debate that was our curatorial focus. We asked ourselves: Whom shall we record? Should we record at all? Can we record the audience? Can individuals choose to opt out?

Before each session began, we made an announcement about the recording; we had previously asked for recording permission from the speakers and panelists. Though most people did not mind being recorded, when someone objected, we respected their will and didn't do so. This was part of the message that we aimed to convey through the festival: that informed consent from citizens to public surveillance is essential.

Recordings from each festival iteration were uploaded to Vidoetage's Vimeo channel.[28] These can be viewed freely by anyone with an Internet connection. Preserving the archive, together with live streaming, reifies the knowledge generated and exchanged. Though the WIKITOPIA 2016 post-event catalogues, with detailed schedule, transcripts and discussions are yet to be published, it will be necessary to make sure that these too are available and accessible for anyone at any time.

Reflecting upon WIKITOPIA 2016, one aspect that could have been improved was in the area of marketing and promotion that would reach out and engage Hong Kong citizens. As a festival designed for social good, we could do more to increase the impact on broad audiences in the city.

Conclusions and Outcomes

WIKITOPIA was inspired by a collaborative mode of online content creation that it seeks to test and apply through offline activities and events. The festival advocates for a free movement of culture while advancing and expanding the boundaries of art. It looks towards a future of knowledge-based self-reliance that results in socially engaged creative outputs.

A great deal of consideration and effort went into cultivating audience participation in open discussions and events at WIKITOPIA. Hong Kong as a site creates certain obstacles that we have had to overcome as curators. These include political pressures and cultural barriers that are at times in opposition to our attempts at abolishing hierarchies of art and discourse.

This process also created opportunities, especially during and after the Umbrella Movement[29] when the younger generation in Hong Kong became more engaged in society, culture, history and politics. Hong Kong youth continue to seek creative channels to express themselves and to participate in broader conversations. One example that emerged in 2011 is Keyboard Frontline,[30] which was formed by a group of active online users who use forums as a base to campaign for democratic values, Internet literacy, privacy and other concerns that resonate with WIKITOPIA's curatorial aims.

Due to a rotation of staff members at Videotage, the organisation of WIKITOPIA has not been continuous or systematised; we are not at the time of this writing certain of when the festival will recur. As a guest curator for the third edition, I am convinced of the urgency and necessity to make WIKITOPIA happen again in Hong Kong. A primary example of this need can be illustrated by a conversation that I had with a journalist friend of mine from Hong Kong. He told me that he travelled to interview and report on a case of injustice in Guangdong, China. There, though he was being very cautious, he was harassed and arrested by the police. This reminded me that his experience could have been different if he had participated in Tactical Technology Collective's workshop at WIKITOPIA 2016, which taught journalists, activists and others about secure communication methods that cannot be intercepted by unwanted parties. It also reminds me of other audiences that could be reached and impacted by the work and knowledge sharing that WIKITOPIA makes possible.

As I look for opportunities to revive WIKITOPIA, my visits in the last two years to the Chaos Communication Congress in Hamburg and Leipzig Germany have inspired me. At CCC, the organisational work is completely carried out by volunteers deemed 'Angels'[31] who contribute greatly to the festival process. In an unequivocally 'wiki' fashion, ideas like these might be applied to generate energy around a new iteration of WIKITOPIA in Hong Kong.

Notes

1 As mentioned in http://bit.ly/2pt0XUc
2 One needs to register as an user in order to contribute to the whole collective body of work on SCP Foundation: http://www.scp-wiki.net/
3 Leuf, Bo, and Ward Cunningham. 2001. *The Wiki Way: Quick Collaboration on the Web*. Boston: Addison-Wesley Professional.
4 Ho, Louis. 2017. "From 'No Cultural Policy' to 'Centralised Market Orientation': The Political Economy of Hong Kong Cultural Policy (1997–2015)." *Global Media and China* 2 (1): 58. journals.sagepub.com/doi/pdf/10.1177/2059436417693007 [Accessed April 7, 2018].
5 Here is an example told by Ai Wei Wei: https://www.nytimes.com/2017/05/06/opinion/sunday/ai-weiwei-how-censorship-works.html
6 Tsui, Enid. 2017. "How Has Art Fared in Hong Kong under China's Rule? The Scene Is Thriving, but Some Artists See – or Fear – Censorship." *South China Morning Post*, June 20, 2017. webcache.googleusercontent.com/search?q=cache:vy-5w2pOI3O0J:www.scmp.com/culture/arts-entertainment/article/2099102/how-has-art-fared-hong-kong-under-chinas-rule-scene+&cd=1&hl=de&ct=clnk&gl=hk [Accessed April 27, 2018].
7 As on their website: hk.creativecommons.org/what-s-cchk
8 Ebersbach, Anja, Markus Glaser, Richard Heigl, and G. Dueck. 2005. *Wiki: Web Collaboration*. Translated by Andrea Adelung. 1st edition. Berlin; New York: Springer, 24.
9 Ibid., 9.
10 Because of the "One Country, Two Systems" policy, it's required for both Hong Kong citizens and mainland Chinese citizens to acquire a special permit to pass any of the 12 control points between Hong Kong and Shenzhen: www.immd.gov.hk/eng/contactus/control_points.html
11 This documentary from WIRED commenting at Shenzhen as "The Silicon Valley of Hardware" can be viewed at www.youtube.com/watch?v=SGJ5cZnoodY.
12 Canaves, Sky, and Juliet Ye. 2009. "Imitation Is the Sincerest Form of Rebellion in China." *The Wall Street Journal, Dow Jones & Company*, January 22, 2009. www.wsj.com/articles/SB123257138952903561 [Accessed March 31, 2018].
13 Mengoni, Luisa. 2015. "From Shenzhen: Shanzhai and the Maker Movement." *Victoria and Albert Museum Blog* (blog). May 24, 2015. www.vam.ac.uk/blog/international-initiatives/from-shenzhen-shanzhai-and-the-maker-movement [Accessed March 31, 2018].
14 The list of current year can be found here: artreview.com/power_100/
15 twitter.com/arthk
16 www.theartnewspaper.com/news/magnus-renfrew-to-launch-new-fair-in-taipei
17 "DIWO – Do It With Others: Resource | www.Furtherfield.Org." http://archive.furtherfield.org/projects/diwo-do-it-others-resource [Accessed February 22, 2018].
18 A catchphrase used by media, one example can be found here: https://www.wired.com/insights/2014/07/data-new-oil-digital-economy/

19 Poitras, Laura, Director. *Citizenfour.* 2014.
20 This report explains the recent controversy: www.sixthtone.com/news/1001996/are-chinese-people-less-sensitive-about-privacy%3F
21 www.dw.com/en/google-street-view-goes-live-in-germany/a-6244398
22 www.ccc.de/en/home
23 accessmyinfo.hk
24 newsroom.fb.com/news/2015/08/connect-with-public-figures-through-live/
25 live.fb.com/about/
26 youth-online.com/wikitopia/live/live.php
27 www.wikitopia.org.hk/
28 vimeo.com/videotage
29 en.wikipedia.org/wiki/Umbrella_Movement
30 kbfl.org
31 events.ccc.de/tag/angels/

17

THE EVOLUTION OF THE ODI DATA AS CULTURE ART PROGRAMME

Julie Freeman and Hannah Redler Hawes

The ODI Begins

In September 2012, the Open Data Institute (ODI) was co-founded by the inventor of the World Wide Web, Sir Professor Tim Berners-Lee, and computer scientist and originator of the interdisciplinary field of Web Science, Sir Professor Nigel Shadbolt, now Principal of Jesus College, Oxford. As an independent, nonprofit, nonpartisan institute, its vision is to help people, organisations and communities to use data to make better decisions and be protected from any harmful impacts.

Since its earliest days, the ODI has had a dedicated contemporary art programme, one of the first art programmes in the world in any organisation to focus on the material qualities and social and cultural impacts of data through art. Berners-Lee and Shadbolt's commitment to open data stems from a belief in best practice in working with data and in Berners-Lee's long-term commitment that his invention is 'for everyone.' It is this ethos that underwrites everything the ODI does – an ethos which resonates highly with media art and maker art strategies concerned with collaboration, participation and democratic engagement with culture.

"Mental aberration" – An Open Call for Art

As an artist working closely with data and interested in the issues surrounding it, Julie Freeman was in early conversations with founding directors and staff about the possibility of involving artists in the new venture. Having invited artist and maker Sophie McDonald to work collaboratively with her (as MzTEK) in 2012, Freeman proposed a project to commission artists working with data to create works for the new office space. As the Institute was initially funded by Government, the decision to commission an art programme did not come unchallenged. However, a compelling argument won: if open data would underwrite the next stage of the World Wide Web, the system increasingly at the heart of many societies' work leisure and socialising activities, artists must be involved in shaping the new networked world order. Introducing the catalogue for *Data as Culture (DaC)*, the inaugural exhibition, which gave the programme its name, Freeman and McDonald wrote:

> Data is driving decisions that shape our daily lives: from friends to governments, we are becoming more reliant on connected statistical data. Global opinion is increasingly communicated through data driven visuals. Personal well-being, sentiment and influence are continually monitored through data harvesting devices. Knowledge at all levels and on all topics can be handed to anyone at any time. Open data is shaping our society.

The ambition, for what would become the ongoing programme, was not only to include work that would not only manifest different data sources but also to work with artists whose practice and projects would challenge our understanding of what data is, and how it might affect and reflect our lives. This ongoing aim of the programme is as evolutionary, slippery and adaptable as the various trajectories of data itself, as it continually takes on new 'forms' in our understanding, vocabularies and infrastructures.

For the first exhibition, the team issued an open call for artists working with data in any way. Within a two-week period, the team received over 80 submissions from over 20 countries. Nine works were selected including Ellie Harrison's humorous and dystopian *Vending Machine* 2009, a repurposed vending machine that donates a packet of crisps each time economic bad news is received from a live feed and *Metrography* 2012 by Bertrand Clerk and Benedikt Groß, which explores how distortions in maps can affect our sense of time and space, as well as projects by Fabio Lattanzi Antinori. Martin John Callanan, Semiconductor and La Société Anonyme. Phil Archer, Stanza and Ben Garrod were commissioned to create new works. Collectively, the works demonstrated the breadth of conceptual and material ways that artists work with data; only one was a screen-based exhibit.

Installing physical works during a new office renovation promoted a thorough sense of integration. Stanza, Lattanzi Antinori and Garrod worked onsite to create and assemble their works, establishing the artists' presence as a commonplace experience in the organisations' early days. The interventions set a precedence.

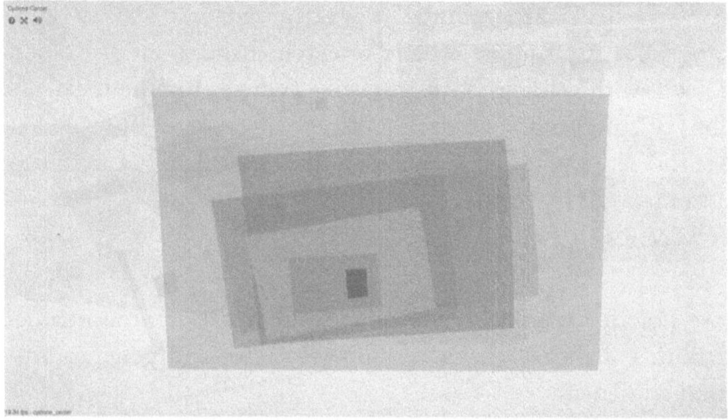

FIGURE 17.1 *We Need Us (Hunt for Galaxy Clusters)*. Image by Julie Freeman.

Data as Culture 2: Open Curating

For the second round of programming, Freeman, who had become the Art Associate of the organisation, put out an open call for curators. In collaboration with curators Honor Harger and Tom Higham, they selected Shiri Shalmy to deliver the second exhibition. *DaC2* 2014 centred around public and private data ownership and access. It critically explored our relationship with surveillance, privacy and personal data. Works included: YoHa's *Invisible Airs* 2011 a series of pneumatic contraptions including a library book being repeatedly stabbed by a kitchen knife driven by open data from Bristol City Council, and Paolo Cirio's *Your Fingerprints on the Artworks are the Artwork Itself*,[1] an online catalogue that uses the cyclical nature of your personal metadata to repurpose other artists' online works into a new artwork, stealing their work and your metadata. Other artists were thickear, James Bridle, Sam Meech and James Brooks. A key curatorial concept for this show was that the ODI headquarters acted as a central hub for the works in the exhibition, with various pieces touring to Lighthouse, Future Everything, the V&A and ODI commercial partners throughout the year – a kind of open-source artwork repository.

Partnerships continued to be formed throughout 2014 when The Space and ODI co-commissioned Freeman to create *We Need Us*,[2] an artwork comprised of a series of animated abstract sound compositions that are dynamically activated by real-time open data. The work echoes the need for the continuing human input essential to the machine-dominated processes of the Internet, reflecting the humanity-focussed aims of the ODI. The *We Need Us* team developed the work on-site, affording staff insight into the development of a long-term art project.

Opening Up the ODI: Artists in Residence

Hannah Redler Hawes formally joined the ODI as Associate Curator in Residence in 2015, creating with Freeman an artist-led curatorial partnership at the helm of the DaC programme. It was agreed that DaC would focus on opening the ODI itself up to artistic research through artist residencies. The theme of Data Anthropologies was conceived addressing the social and cultural implications of emerging open data landscapes and placing people firmly at the centre. Artists were invited to look closely at current habits and future trends in behaviour and social interaction.

Artists were researched

- Whose practice would be relevant to our main communities of interest; ODI staff, collaborators, clients and associates and general, arts and tech cultural audiences;
- Who have the personal qualities to contribute to and integrate with ODI, its team and its activities as appropriate, through discussion, dialogue and/or collaboration and public-facing events;
- Whose existing work would lend itself to an on-site exhibition.

It felt essential to keep up the momentum of placing art in the working environment. By showing existing work whilst in residence, the artists provided those with whom they would be interacting with a sense of who they were through their art. In addition, it was important to situate the DaC programme in a continuum of artistic practice that has connections to broader arts histories.

Contemporary art duo Thomson and Craighead and photographic artist Natasha Caruana were directly approached to be ODI artists in residence, due to the ethnographic approach of each and their ongoing and subtle uses of data in different ways. Thomson and Craighead offer deceptively gentle critical explorations of the structures and social constructs of the Internet and social media spaces. Caruana's work is concerned with narratives of love, betrayal and fantasy, drawing from archives, the Internet and personal narratives. Her research sources naturally raise questions around how today's technologies impact relationships. The residencies would explore how Thomson and Craighead's 'armchair anthropology'[3] and Caruana's bold, inventive uses of performative strategies and scientific processes develop the open data conversation.

The artists spent six months in the organisation where they were given free rein to relevant research and access to all staff at every level. Some conversations were engineered; others emerged organically. They were asked to be as open as possible throughout the process, and joined staff meetings, including daily stand-up, where everyone states goals for the day. During this time, a successful Arts Council England funding application enabled the artists' proposed new works to be commissioned. These were to be premiered in a co-curated project developed with FACT Liverpool titled *The New Observatory.*[4]

Craft, Code and Sound

In 2016 the 'in residence' approach continued through a partnership with new music charity Sound and Music and their embedded programme. Following an open call, artist, musician and programmer Alex McLean was selected as the Embedded Sound Artist in Residence. McLean is a pioneer of live coding and algoraves (algorithmic raves) – sound and visual performances generated by computer code that are created collaboratively in real time. The working environment was not conducive to an exhibition of McLean's performance-led sound work; however, looking more deeply into his practice, his collaborative nature was striking. McLean works with a wide variety of other artists, coders, makers and designers, whose practices share a strong sense of openness where the process of making is prioritised over any finished artefact.

Redler Hawes and McLean collaborated to co-curate *Thinking Out Loud*, which drew connections between the ways in which humans have captured, encoded and distributed data, and have, historically, made it meaningful through pattern. It featured makers who are driven by radical intentions to expose the inner workings of the systemic structures we live in, through works such as Pre-Columbian quipu, computer-coded forkbombs and the sounds of the Shetland Islands.[5] Few of the works in the exhibition related directly to data, yet the ethos of process, collaboration and co-production reflected and amplified the positive messages behind ODI's championship of an open data movement. *Thinking Out Loud* also revised 'macho' stereotypes of technology as sleek, finished and impenetrable.

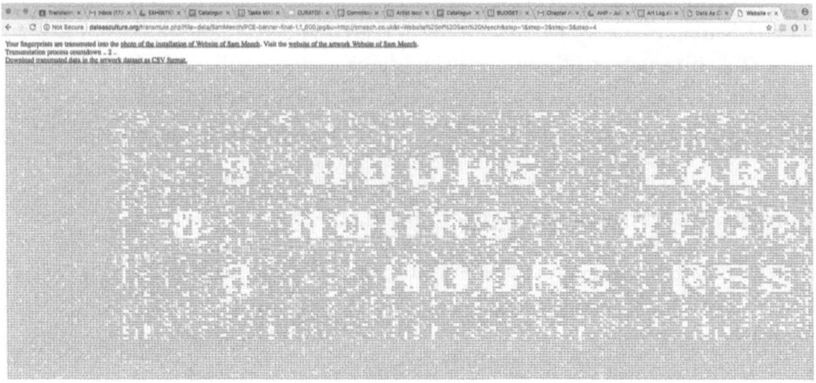

FIGURE 17.2 *The Fingerprints on the Artworks are the Artwork Itself* (Sam Meech detail). Image by Paolo Cirio / ODI.

Observatories of Ourselves

In 2015, using the proposals for new works from Thomson and Craighead (*Recruitment Gone Wrong*) and Caruana (*Divorce Index / Curtain of Broken Dreams*), along with Freeman's *RAT.systems,* a collaborative external curatorial project was sought with external organisations. Mike Stubbs, Director of FACT Liverpool, proposed we work with their PhD candidate Sam Skinner whose research addresses Liverpool's history of observatories.

The New Observatory, co-curated by Redler Hawes and Skinner was the major summer exhibition at FACT in 2017. It convened an international group of artists, researchers and technologists whose work explores new and alternative modes of measuring, predicting and sensing the world today, areas all heavily imbued with data. The exhibition considers how conversations around observation shift when we relocate the traditional purpose of an observatory from a specialist site of scientific exploration to a free-form site of artistic and individual exploration, exempt from formal constraints of science. Key to the concept is that imagination and speculation are as fundamental to these activities as any technological tool. The exhibition positioned observation as a proactive occupation including drawing, composing, film-making, expeditions and citizen sensing. Artists included Wafaa Bilal, Jackie Karuti, David Gauthier, Evan Roth and Citizen Sense.[6]

Commercial Collaboration

2017 saw a curatorial collaboration in a corporate context. As part of a relationship between the ODI and the Digital Catapult, Redler Hawes curated *Hybrid Landscapes* for their headquarters with 11 artists who respond to and subvert digital technologies in surprising and unexpected ways.[7] Placed in the locus of the UK's major centre for developing research and technologies around smart cities, the Internet of Things, privacy, trust and surveillance, the artists presented works that explore how lived experience plays out simultaneously across natural, built and networked worlds, building new perceptions and perspectives.

Laughing All the Way to the Data Bank

2018 trialled the use of an emoji as a title, representing something highly communicative that remains unsaid. The *LMAO* 😹 open call invited artists to consider

> how in these meme-fuelled, statistically 'mythological' times, data, and the algorithms that thrive on it, are often presented as a privacy-obliterating risk-based menace; and how there are always two sides to a story and that data can also be a force for good, as well as game for a laugh.

The premise being that humour helps us reveal failures and cracks in the system. Works were selected for their playful yet critical approach to data and its uses. Irreverent, provocative, unconventional and plain silly, they asked us to challenge our preconceptions of data, and consider the humanity behind our technologies. Participating artists[8] poke fun at the ineptitude of Google's image search capabilities or the expectation that 'big data' will predict the future, persuade us that sharing poop is a Good Thing and place a taxidermied cat in the ODI ceiling in homage to the Internet's most famous visual meme.

Reflections on Success and Lessons Learnt

Data as Culture forms and asks new questions and creates a visually stimulating and intellectually rich environment. The exhibitions increase the awareness of what data is and how it can be used creatively, and they help us to 'realise' some of its more abstract qualities. Many of the works critically problematise conversations around the opportunities and concerns data presents, such as construction, navigation and dissemination of new forms of knowledge, and the balance between transparency, trust and privacy. Crucially for a constantly learning innovation company, DaC shapes and influences the culture of the organisation: continually challenging its established order literally with change-over exhibitions, and ideologically through the challenges new works bring. DaC reaches, communicates and engages with new, unexpected and diverse audiences – thousands have visited the exhibitions and millions have been exposed to the work through broadcast and social media.

Internal feedback has demonstrated that staff feel valued by an organisation that brings artists in to work alongside them and puts museum-standard work on the walls. They appreciate the intervention of art and artists that creates an exciting and dynamic working environment allowing for, and encouraging, play and lateral thinking.

Staff actively engage with the way the art provokes debate on key societal and ethical issues they feel are pertinent to their work such as censorship, online behaviour and the political ramifications of data.

Audience and the Importance of Partnership

The immediate audience who visit the ODI are policy makers, trainers, international development specialists, researchers, technologists, politicians, start-ups, multinational CEOs, international delegations and civil servants. They visit for training, meetings or programmed talks. The art offers a more enjoyable place for training, adding an extra dimension through unexpected informal learning, bonding and discussion. Trainers often use the artworks as ways to start conversations about complex data-related issues.

The exhibitions attract an arts audience including artists and curators, galleries, creative industries and students, for whom bespoke tours or seminars are available. Members of the public can make an appointment to view the art; however, we reach the widest general and arts-specific audiences through our partnerships with other galleries, festivals, conferences and corporate events. *The New Observatory* reached 6.5M people through broadcast and social media channels and over 50,000 visitors, 16,000 of whom directly visited the gallery-based exhibits.

Journalists are enthusiastic about the storylines created by the programme, and it has been covered in *The Guardian, The Telegraph, New Scientist, Wall Street Journal, Wired, PostMatter, Motherboard, TechCrunch, BBC Click, Mashable* and more. Specialist arts press have been less engaged, which perhaps demonstrates the difficulty of finding a 'critical' home for interdisciplinary practices in art-making and curating in this context.

Mediating the Encounter of Art Beyond the Gallery

We do not bring art into the office space to achieve consensus. Debate and difference of opinion are important aspects of engagement. A degree of disruption and discomfort can act as agitators to prevent staff from gravitating towards obvious solutions. Bringing art into a working environment is different to placing it into a cultural destination where visitors are free to make choices about what they experience. In a working environment, staff do not always have a choice of where they work, or how proximate they are to an artwork. Highly personal readings are formed through 'living' with work, which deepens with regular coexistence. This can be intimate and intense and, in some instances, can generate discomfort. It is difficult to anticipate the breadth of visual and emotional triggers – some relate to prior experience, others to sensitivities in wider culture.

One work which included humorous attempts by people on YouTube to hold their breath under water was removed due to a member of staff having experienced a related family tragedy. Another work which broadcast commonly used search terms started off amusing staff, but there were times when terms became deeply offensive, obscene and misogynistic due to the artists' uncensored approach. The critical context of the art gallery allows an artist to hold an objective mirror to online behaviour, however in the context of a working environment, the inability of staff to move away from such content renders its inclusion untenable. There was unanimous agreement that it would be impossible to continue displaying the work with such obscenities. Even when an artwork presents unexpected problems, the way the team collectively responds creates dialogue and a valuable shared experience.

In addition to issues of censorship, enthusiasm for DaC is not always universal. When the economic environment is challenging, not everyone recognises the value of the art programme or they can recognise the value but do not feel it should be a priority. Constructive criticism and feedback are invited and staff provocations help shape the programme and how interpretation strategies are developed.

Artists' Responses

Artists are proud to be part of the ODI story. They appreciate contributing to a project led by eminent global leaders like Berners-Lee and Shadbolt. Although we are committed to high-quality museum-standard installations, ODI is not a gallery. Artists are unanimously generous in understanding the alternative context of our specific setting and of the relationships visitors will build with them and their work. Many of the artists contribute to the free ODI Friday lunchtime lectures, which are streamed online and available as an archive. Here, artists are included alongside government officials, cultural and business leaders, major museums and other influencers.

We encourage artists to evolve their practice through their work with ODI. McLean felt he expanded his curatorial practice through the collaboration as well as creating a new composition funded by Sound and Music; Thomson and Craighead set up an internal research blog as a place where they could discuss ongoing research. They had never worked in this way before stating '*This is a method we will use in the future as it helped us engage with our early research in a more critical manner.*' Caruana used her residency and commission to move beyond photographic works into installation. The curatorial support of the DaC team provided her with the necessary confidence. In a post-residency self-reflection piece, she writes,

> To generate research from the conversations, observations and critic of an office environment gave my practice a wealth of material to work with. I found the process of exhibiting in an office environment an extension to my core belief that art should be for all. To take my work into working environments was a way to break out from conventional exhibition spaces and make contact with a different audience.

Future Directions Informed by Risk

DaC works with established and emerging contemporary artists, familiar and new to working with data. Its aim is to ensure that dominant narratives are continually challenged and that data futures are inclusive and representative of a wide range of citizens' issues and views. The programme makes data accessible and visible to new and different audiences. It offers a unique access point to reflect on how the data revolution might 'affect me personally' and offers a progressive encounter with open data, art and technology to broader audiences.

This reinforces the ODI mission to affect societal and cultural change with how data is used.

An awareness of a concerning lack of diversity in our programming has opened private and public conversations with leading artists of colour to consider how to address this. Deborah Williams, Gary Stewart and Derek Richards worked with us to shape four key questions to consider at the 2016 Summit and which could be put to the entire data community: How can we write multicultural histories of data? Who controls the sources of our data? How can data infrastructures define culture and what is the cultural impact when data fails? These inform the forward programme which includes for the first time a much deeper involvement with core ODI projects by looking to see how artistic strategies and disruptions might influence and be part of technical research and development projects and the telling of data stories. Propelled by informed chances, curiosity and creative risk management; the programmes' dynamic evolution is an exciting adventure to be part of.

> The Data as Culture programme helps to bridge the gap between the material world and the digital, by giving intriguing and tangible examples that demonstrate how data is embedded in our everyday lives.
>
> *James Maddison, staff member, ODI*

Notes

1 http://www.dataasculture.org
2 http://weneedus.org
3 Cook, Sarah, 2013. In *Thomson & Craighead – Flat Earth* by Thomson & Craighead. Dundee Contemporary Arts / MEWO Kunsthalle.
4 22 June 2017–1 October 2017.
5 Featuring Alex McLean, Dan Hett, Felicity Ford, Antonio Roberts, Ellen Harlizius-Klück, David Littler, Sam Meech, Amy Twigger Holroyd and David Griffiths & Julian Rohrhuber.
6 Also: Burak Arikan, James Coupe, Phil Coy, Interaction Research Studio, Rachel Jacobs, Kei Kreutler, Libre Space Foundation, Stanza, Liz Orton, Proboscis, Jeronimo Voss, and Yu-Chen Wang.
7 Katriona Beales, Daniel Brown, Lewis Bush, Field, felicity Hammond, Manu Luksch, Antonio Roberts, Evan Roth, Thomson & Craighead, Suzanne Treister, Richard Wright.
8 Caitlin Foley & Misha Rabinovich, Ellie Harrison, Dan Hett, Franco and Eva Mattes, Lee Montgomery, Riitta Oittinen, and Pip Thornton.

18

TRACKING HACK-STYLE INTERDISCIPLINARY PROCESSES AT LABORATORIO ARTE ALAMEDA, MEXICO CITY

Tania Aedo

Laboratorio Arte Alameda (LAA)[1] is a space where intersections between art and a multiplicity of disciplines are at the core of our inquiry. The museum's unique building is a temple, a remain, a fragment of the Ex Convent of San Diego. On one street is Dr. Mora #7, completely renewed recently along with Alameda Central, one of Mexico's most important public sites. On the other street is Colón # 5, a building constructed as a result of the earthquake of 1985, designed by architect José Luis Benjure, where our Documentation Center CDPL is located. This building is back-to-back and sharing space with Diego Rivera's *Museo* mural and built over the terrain where Hotel Regis stood before the 1985 earthquake, where Diego Rivera's rescued mural *Sueño de una tarde dominical en la Alameda Central* was originally painted. For LAA, the relatively recent activation, or re-activation, of *maker, hack* and *DIY* cultures began to revitalize questions as central as art *practice* itself. *Making* and *doing* are often identified with craftsmanship and placed in contrast, or opposition, to thinking or theorizing. These types of categorization are among the processes under constant inquiry in curatorial and museum practices, especially when they cross boundaries of disciplines or realms.

Apart from purely conceptual artists – who are quite a rare species – all artists are makers, and in the end, so are all people. Nothing compares to the feeling of inventing something and constructing it; it is in our nature. The act of making, however, has gradually been taken away from the hands of everyday people while the role of the 'maker' has been assigned and reduced exclusively to some artists, artisans and amateurs working with *arts and crafts*. This further emphasizes stereotypes and dichotomies such as the distinction between mind and body. The capitalist drive to homogenize culture has distilled the workspace of most western, wealthy and middle-class people, who are trained to keep working on computers in a simulated digital environment that uses the most mundane desk metaphor for its interface. Typing has become the handiest activity undertaken in most people's daily lives.

In the meantime, the stability of the planet has reached a critical point that is calling for everyone's participation. Sustainability must be a central concern if our species wants to continue to populate this world and collaboration is necessary to include everyone in this plan. But how can we re-learn how to work together? Artists – like everyone else – must ask and seek answers. Some of them are doing this very well by testing forms of collaboration, investigating key topics across a plethora of disciplines, experimenting with prototypes for environmental sustainability, building, hacking and presenting models for change.

Like critical thresholds that are passed when a paradigm shift occurs, abundant knowledge can be generated when we cross disciplinary boundaries. In order to track, trace and share these processes, experimental documentation methodologies are needed. These are among the key issues that may be observed and experienced through the interdisciplinary collaborative artistic practices happening at LAA. At the same time, these concerns benefit from an increasingly *maker, open* and *hack-style* approach.

At LAA, we ask ourselves: How do audiences, artists, researchers, students, parents, teachers, LAA's guards, technicians and curators remember, record or document their experiences at our space? How do they narrate those experiences? We work to motivate DIY archiving and preservation practices. These include workshops on bookbinding and preservation of digital formats with non-specialized technologies, such as the workshop 'Preservar el tiempo' by Ollin Yahvé. This was directed at amateur or non-specialized audiences interested in the preservation of personal and family archives.[2] We do this because our experience has shown that *amateur* practices are important. We believe they are key to the future preservation of contemporary culture and for addressing many other future challenges. The documentation of so-called amateur practices becomes an important source for our own research because it generates knowledge that is ready to be opened and shared.

Curatorial practice at a space such as LAA offers an arena for hands-on meditative collaborative processes. As we respond to a hybrid scene with a wide set of interests and multiple titles (i.e. electronic or media art, sound art, net. art, conceptual art, media archaeology, expanded cinema, movement art, new materialisms, art/science/technology practices, experimental electronic music, critical theory and so on), we sense a tendency and a necessity for openness and collaboration in order to deal with these wide universes and to be able to share the knowledge generated through them.

What kind of knowledge is generated at a place like LAA? How do maker, DIY and hack-style practices support our approach to knowledge generation? We use a variety of methods, including educational programs that may be replicated to create exponential impact, exhibition design experiences with a focus on historic buildings – instructions, plans, non-invasive and sustainable methods – and shared experiences on archiving and preservation. The results of our research in these areas may be useful to other curators and institutions.

LAA has traditionally been a site for art projects that challenge perception in many ways.[3] Here, sound art as a practice – one that directly shakes the somehow still present idea of art as a mainly visual phenomenon – has emerged as a key area of inquiry. Phenomena of perception as they are closely linked to knowledge, in life and in the exhibition space, have been explored in many of our endeavors as workshops. Our 2016 summer program for children, for example, focused on the topic of perception. We have also been tracking the idea of perception in our seminar on curatorial and heritage theory, and there are many other examples.

Our recent project, *Tianguis de conocimiento*, leads us to the setting of our building's threshold. Using the term *tianguis*, the Náhuatl (still used) word for small markets, with the concept of trueque (barter), this project aims to offer everyone, from our audiences to passersby, a quick hands-on opportunity to experiment with an open-access artistic production-experimentation site. Situated directly in the threshold between the street and the museum in LAA's Atrium, *Tianguis de conocimiento* is fueled by hack-style, DIY, maker approaches that allow participants to 'put their knowledge on the table,' have conversations on and around it, interchange it with the knowledge of others and in the end have something different that is continually emerging. At the same time, the project raises the question – what does *knowing* mean? We like to think, as Francisco Varela, the Chilean neuroscientist, that learning is just a change in what we know,[4] and as Humberto Maturana, biologist, also Chilean, that knowledge is a key ingredient of life. We *know* with all our cells that *living* is mostly the same as *knowing*. The famous saying by Maturana, 'todo conocer es hacer y todo hacer es conocer,'[5] is what we aim put into play with *Tianguis del conocimiento*. Making and DIY practices frequently remind us that it is important not to organize knowledge in a hierarchical fashion in which knowing 'what' is of higher importance than knowing 'how.'

LAA is a space that offers a constant flow of public programs beyond its exhibitions: live acts, conversations, workshops, concerts, seminars, festivals and many other formats. We are open to a multiplicity of subjects and conversations around contemporary cultures. Through our programs, we frequently engage with artists and groups whose methods align closely with hackstyle, DIY and maker practices. As a curator, as museum director, and as programmer, it is important for me to seek artists or collectives who are flexible. They must be willing to integrate with our local crew, learn and teach by doing and collaborate as active agents in issues related to the environment and labor, such as museum workers' schedules, taking care of how exhibition debris will be used after a show closes and caring for the well-being of the guards while they look after their artworks (this is particularly important when we talk about pervasive sound, audiovisual or installation artworks which perform continuously for eight hours, for example). These types of practices, then, can be viewed as possible ways out of a hierarchical order toward more horizontal, ethical and environmental friendly collaborations in the museum context.

Following, I will profile some case studies of collaborative, DIY, interdisciplinary guest agents and projects that have been presented at LAA. Next, we will briefly examine a set of hackstyle strategies staged by curator-artist-producer, and museum-hacker Paloma Oliveira de Andrade in three parts: an exhibition, *Transmutación: Alquimias del espacio* (Transmutation: Space Alchemies); the leading of a committee for a friendlier and more accessible museum, and the constant animation of our sustainability program in collaboration with our environmental partner, Fundación INBA.[6]

Colectivo Chipotle

https://colectivochipotle.org/ gathers visual artists, programmers, cultural producers and musicians, who share an interest and experience in video games, informatics aesthetics and 8-bit music. Chipotle regularly organizes chiptunes and circuit bending workshops. The collective also organizes *Format DF*, the first annual international chiptune festival in Mexico City. This open-access festival offers to many young musicians and enthusiasts experiences centered around experimentation with 8-bit sound as well as hacking and circuit bending old toys.

Astrovandalistas

http://www.astrovandalistas.cc/base/ was founded in 2010 in Tijuana, a city in Northern Mexico on the border with the United States. Tijuana is mostly a city of 'maquiladoras,' in which large-scale industries such as electronics, aeronautics and energy have established thousands of assembly plants. Tijuana's industrial reality and geographic location give it a particular relationship with technology. Astrovandalistas respond to this by combining research, artistic action and experimentation with different technologies, activism, urban hacking and open knowledge. They operate as a decentralized laboratory in different locations in Latin America, such as Mexico and Brazil as well as in the United States, developing low-tech tools that enable new forms of communication and collective participation with open software and open hardware that can be easily replicated.

MusicMakers Hacklab

http://mmhl.org/mx/

MusicMakers Hacklab is a week-long open collaborative laboratory focused on the exploration of possible relationships between body and sound, hosted by Peter Kirn of CTM Berlin and Leslie García/Paloma López in Mexico City and Tijuana. During the MusicMakers Hacklab week, each of the participants builds systems and instruments based on processes such as sonification of biological organisms, aural and sonic stimulation and the amplification of body sounds. At the end of the laboratory, participants show their projects as live acts. MusicMakers Hacklab has released and recorded pieces in cassette editions alongside a printed publication. The questions and explorations proposed by these collaborative and open laboratory dynamics trigger singular experiences that enable us to see the importance and potential of artistic research in understanding and inquiring into our present.

Interspecifics

http://interspecifics.cc/

'... is a multispecies collective experimenting in the intersection between art, science and technology, which embraces hybridized practices among different disciplines and living organisms, open knowledge and precarity as a challenge.'[7] They are focused on exploring hybrid systems that compound bio-organisms and machines. Their projects include *The Energy Bending Lab*, an instrument comprised of a set of custom-built modular synthesizers and transduction tools that create real-time sonification from the electric properties found in bacteria. They use cymatics – visible frequency vibrations – that are converted into a visual pattern generator. Interspecifics' DIY approach results in an effective method to share and open up what would otherwise be closed knowledge.

At LAA, we have seen how so-called 'general audiences' can gain new understanding of the spectrum between living, non-living and the fuzzy border between these, which are key to understanding our present and exploring what life may be like in the future. Audiences are not only able to listen to or read about crucial developments in scientific knowledge related to Artificial Intelligence/Machine Learning or Neural Networks, but they are invited be hands-on with the work. For example, experimenting with microorganisms that are capable of learning and composing music displaces the idea of a human-centered culture. Interspecifics' continuous sharing of experimental and artistic practices and methodologies is what makes this process so exceptional and a very different experience from learning about synthetic biology through regular formal instruction methods.

One of the recent projects by Leslie García and Paloma López, the two nodal members of Interspecifics, is *Speculative Communications*.[8] This project focuses on learning from interactions among microorganisms while creating the artificially produced sapience of the composer as a live sound and visual act. There is potential for theoretical and artistic speculation, as the name of the project, *Speculative Communications*, states. There is a tendency to generalize all processes such as this as 'sonification' and consider it a reduction of microorganisms' behaviors with no significant artistic merit. If we let ourselves go deeper, however, in understanding methods of communication or decision-making among creatures or organisms that are present in this artistic speculation, we may consider conversations that reflect our present, such as those around consciousness. Through material speculation and design, Interspecifics' work is always full of references to timely topics. In the case of *Speculative Communications,* these relate to ideas that Humberto Maturana investigates in his laboratory in Chile. Interspecifics are among other Latin American artist groups who closely read key theoretical figures such as Karen Barad, Gilbert Simondon and Humberto Maturana and reflect their thinking through making in the form of experimental laboratories and knowledge-sharing practices. In many ways, this collective, as well as individual works by Leslie and Paloma,[9] are remarkable examples of proposing models for a better world. We can use speculation, sensing and questioning in the museum space to draw out rich discussions, but also, importantly, to help us imagine possible futures.

***Transmutación: Alquimias del espacio* (Transmutation: Space Alchemies)** is an exhibition and curatorial project by Paloma Oliveira and artist Jaime Lobato specifically conceived for LAA in close dialogue with the historic building and with our mission. This project connects different modes of conceiving knowledge and perception with our environmental and inclusion[10] goals. As with other projects supported for the last two years, *Transmutación: Alquimas del espacio* focused on perception. We began to notice that many of our exhibition and live-act projects were challenging people's sense of self and place, guiding what they would listen to or see and what they would filter out. Perception in the *Transmutación* exhibition is seen as a political act, put in practice as a dialogue between curator Paloma Oliveira's background in art, museology practices and production and Critical Disability Studies in tandem with the artist Jaime Lobato's knowledge of sound and space, archaeoacoustics, and echo-localization. Both artist and curator are complete hybrids in their activation of knowledge through practice. Critical Disability Studies, in the words of Oliveira in the printed brochure of the exhibition, challenges the ablest assumptions that shape society. It helps us to become conscious about everyday power structures, particularly those that we are not accustomed to perceiving. This extends to the fact that art as an institution is still to a great extent based in the *visual,* a mode of perception that is not available to all people. The *Transmutación* exhibition subtly but decidedly questions many prejudices on the 'normality' of bodies and their behaviors. This functions through the artworks, but is also integral to the museography, including accessible new infrastructures and the welcome and communication the museum offers to visitors. Because of the unique architecture of LAA, sustainability means also considering historical heritage as living space. *Transmutación* has been included in a set of experiments and inquiries on sustainable development (based on ONU sustainable development goals[11]), inclusion, and research conducted on historical heritage buildings that are inhabited by contemporary art projects. This includes recycled museography in collaboration with Juan Serrano, carpenter, who was willing to build a 24-square meter pyramid out of wood that was not-standard size and utilized reused materials while also rebuilding every piece of furniture in the exhibition. We also used only LED low-consumption lights, standalone furniture and light vinyl for the installation of *Transmutación*. This is a method that we plan to replicate in future exhibitions.

Conclusion

At LAA, problematizing knowledge is intertwined with our mission. As a matter altered by historical and epistemic violence, *knowledge* indeed means something different to different people. Physically, LAA is grounded in the historical. We inhabit an ex-convent from the 16th century, a time when people thinking differently would have been burned – and there is a small plate in our entrance to remind us of this fact! In a place as rich in traditional knowledge as Mexico City, all of that sapience needs to be re-encountered and put into play with what is considered *formal* knowledge. As a result of such a mixture of cultural specificities, what we provisionally conceive of as *knowledge* needs constant investigation, along with many other subjects as enormous and problematic as this. At the same time, it is important that we stage our questions sooner rather than later, and that we learn from that staging in creative and effective ways.

Museums and contemporary art exhibition spaces always have more to learn. The use of live organisms in our programming, for example, has presented a big challenge. When using live organisms of any scale, from microorganisms to plants, it must be decided how to preserve, maintain or consider disposing of them. This is something that can be documented and shared with other museums or galleries.

Through our methods of working, we have seen an extension of the relationship between audiences and exhibitions. Maker, DIY and hackstyle approaches consider what people are taking home after the exhibition or event – whether in terms of experiences or knowledge-sharing. Audiences are sensitive to this, and they come back looking for more of these kinds of interactions. As we have discovered this as an institution, we have learned to include museum-hackers on our team and take on some of these strategies ourselves as we assume this museum-hacker position. The joy of this learning is what I have aimed to describe, although very briefly, through this text.

The learning that we do as an institution always take place alongside and within practices of making, hacking and doing. Therefore, I will conclude by discussing a cascade of projects that are currently being rolled out. I am finishing this essay while *Tianguis del conocimiento* has hosted two more projects. The first is a curatorial series by Gemma Arguello's called *Emplazamientos de la memoria*, which gathers together a group of artistically relevant and diverse projects. In this case, Gemma is not part of the LAA staff but is an active member of the contemporary art scene and conducts academic research that is critically oriented and emphasizes action in the public space. The series of interventions in *Emplazamientos de la memoria* are about memory in relation to Alameda Central, a location that is key to the configuration and understanding of our enormous city. Second is the project *La lengua del diablo* by a collective from Tlaxcala called *Arte a 360 grados*, about the extinction of languages, in this case Náhuatl. At the same time, we have been increasingly accessible in terms of mobility. This is demonstrated as we open *Arqueologías de la autonomía*, which challenges ideas about disabilities and perception as means to explore ways for our species to better understand one another and live together.

Notes

1 Laboratorio Arte Alameda, LAA, is an exhibition space and a Documentation Center, CDPL (Centro de Documentación Príamo Lozada, named after our founder curator), and is located in downtown Mexico City. LAA inhabits the remains of an ex convent dated from 16th century: Ex Convento de San Diego. Part of Instituto Nacional de Bellas Artes (National Institute of Fine Arts) INBA, its core interests are the intersections among disciplines, sapiences, realms, especially those among art, science and technologies.

2 https://www.cultura.gob.mx/recursos/sala_prensa/fotogaleria/activacion_cdpl.jpg

3 Examples of projects that have addressed perception include: *Primer Movimiento* by Tania Solomonoff, 2017, *Zee*, by Kurt Hentschlager, 2013 and *SIN* by Mario de Vega, 2014.

4 Maturana, Humberto R., and Francisco J. Varela. 2004. *El arbol del conocimiento*. 1st edition. Santiago de Chile; Buenos Aires: Lumen.

5 Ibid., 13.

6 An ONG helping the development of the National Institute of Fine Arts INBA's institutions, as museums.

7 http://interspecifics.cc/work/

8 A project developed at Medialab Prado, and the National University in Santiago de Chile and partly funded by Laboratorio Arte Alameda.

9 Along with other artists working closely to hackstyle, maker and DIY practices of sharing knowledge and promoting engagement with complex systems and with *matters of concern*.

10 The words inclusion and disability are also under question always as we are working towards the depathologization of difference and genuine openness in our institution.

11 https://www.un.org/sustainabledevelopment/sustainable-development-goals/

19

THE MAKING OF DIGITAL FUTURES

Irini Papadimitriou

> At a meta-level, Critical Making aims to focus attention of the ways in which materially-engaged activities provide cognitive resources for thinking through complex individual, social, and societal issues. In other words, critical making is an elision of two typically disconnected mode of engagement in the world – 'critical thinking', often considered as abstract, explicit, linguistically-based, internal and cognitively individualistic; and 'making', typically understood as material, tacit, embodied, external and community-oriented.
>
> *Matt Ratto and Stephen Hockema*[1]

In 2008, the year Matt Ratto coined the term 'critical making,' we launched Digital Programmes as part of the Learning Department at the Victoria & Albert Museum. Ideas such as critical making, exchange and critical thinking, as well as investigating the role of the 21st century museum became central to our work. In the next few pages, I will share part of this journey, from exploring digital learning tools to attempting to transform the Museum into a test bed for exploratory cross-disciplinary activity through the delivery of labs, workshops, talks, installations, hacks, making, gaming and more.

Before 2008, and similar to colleagues in other art institutions, we had already been exploring the use of Web 2.0 in the context of a museum environment;[2] we had been experimenting with online platforms, from social media to blogs, podcasts/vodcasts, wikis, apps and others, to share content and engage with audiences in new ways. Web 2.0 had enabled museums to reach out to wider audiences, inviting them to connect with collections, to 'curate' objects and participate in discussions online, but what we were also interested in was how these online tools could benefit our work with visitors in the physical space.

We were starting a new programme at an exciting time, when accessible and easy-to-use technologies were at reach, tools that could enable collaboration and participation at a large scale, as well as access to resources and sharing knowledge and skills with communities. But if the purpose of our team was to create a space to help engage people with Museum objects and inspire them to learn, be creative, and make with digital tools, then we had to consider the 'digital' beyond just referring to the use of technology.

We couldn't ignore the fact that, being at the V&A, we were surrounded by collections that span over 5,000 years and included virtually every art form, from sculpture, painting, architecture and works on paper to textiles, ceramics and performance. An important part of our work was to show the relevance of the digital to all these art forms. One of the most common things I had to explain people in the early days was how technology and digital don't imply only screen-based content.

Most importantly, we wanted to explore and question what is digital, where it exists, how it shapes our lives and the world today, how it might exist and change our world in the future and who has control over these decisions.

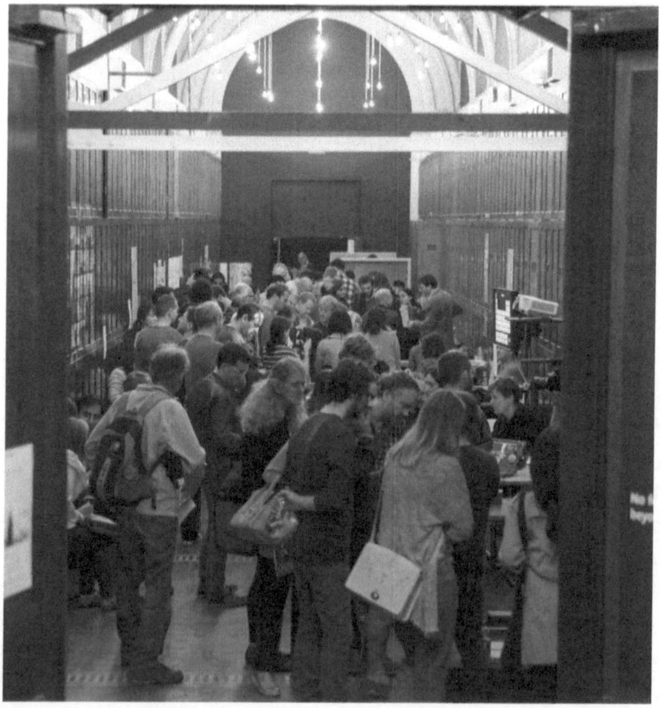

FIGURE 19.1 Digital Design Weekend Open Collaborative Lab. © Victoria and Albert Museum.

We had the opportunity to build programmes and events that not only presented innovative work but also focussed on process-sharing and research that could motivate others to engage with digital art and design at a different level. These process-sharing and exchange events developed to include tinkering and making sessions or hackathons (hacks), where individuals with different skills come together in a design/making sprint-like activity to create collaborative work, mostly responding to a challenge. The hackathon format has been used by quite a few art institutions as a way of bringing together attendees from different communities such as software developers, engineers, computer programmers, games designers, graphic designers, but also scientists, artists and makers.

Here, I share some of the hack activities and events we developed in our programme and how we experimented with formats in order to reach more people and make them more inclusive focusing on projects that combined technology with traditional making skills, but also conversations, in order to enable wider audiences to participate. Allowing space for conversation was a significant part of these sessions, as we imagined them not only as a place to build and complete a project, but as process-based events, where attendees could exchange skills, teach and learn from others, leading to further exchanges between them in the future.

In a hyper-connected world, we are constantly exposed to technology and digital; on one hand we have tools that can give us a voice, new ways to exchange knowledge, take action and influence change. Technological developments of unprecedented speed have also opened the way to great advancements in the fields of health, design and engineering, to name a few. However, although technology is inseparable from everyday life, for most of us, this relationship with the digital is a superficial and consumerist one. We don't understand the complexity of our technologies or what lies beneath our everyday devices; from how and where they are made, or end up, to social implications and ethical issues hidden behind; mass surveillance, algorithmic control, obsolescence and environmental issues, as well as use of conflict minerals and working conditions. So now, more than ever, we need to reflect on the role of the digital and take a critical stance about the complex issues involved.

When it comes to our digital programmes, from co-design, prototyping and collaborative making sessions, an important objective for these events was to think through ways in which we relate to the world, to stimulate critical thinking and to place technology in the context of society and current challenges. At the same time, we wanted to focus on taking control of many of the tools that surround us, mend and invent. Could we, for instance, with productive collaboration, openness and more inclusion, create more human solutions for our environments, communities or cities? Could we imagine more human visions of the future with common points of reference and shared values?

It is impossible to talk about the V&A's Digital Programmes without mentioning making and the maker subculture. Our work was developing about the same time when, in the UK, there was continuous rise of makerspaces, fablabs and hackspaces,[3] where makers, hackers, technologists, artists and other like-minded people could create, share work and get support for their projects. Such spaces and communities have been closely aligned with the DIY, hacker ethic and open source movement of the sixties and seventies. A combination of the Web with accessible, affordable technologies, desktop manufacturing tools and traditional making techniques opened up endless possibilities for contemporary maker culture.

Although our goal was never set to create any form of makerspace, since the beginning our programmes focussed on open source and a shared learning approach, while exploring intersections of art, design and craft, it felt appropriate to engage with the maker and open source community.

Our first attempt to work with such a community at the Museum was with Hellicar & Lewis[4] and the openFrameworks Lab. The event, which was something between an open workshop and a hack, brought together over 50 people from the openFrameworks community across the world, who responded to an open call to develop a series of interactive projects with input from Museum visitors. Developed pieces included public projections to encourage pedestrian interaction in public spaces, an installation generating silhouette portraits from visitor data, as well as other human computer interaction and computer vision projects.

openFramerworks Lab took place soon after Decode: Digital Design Sensations,[5] a major exhibition at the V&A that explored how digital technologies provide new tools for artists and designers, and which offered us the chance to work closely with many artists, whose work merges code, design, art and craft skills. The event also happened to be part of the V&A's first Digital Design Weekend,[6] an annual event that was initiated to open dialogue between creative practitioners, create a big platform in the Museum to present cutting edge work and intersections in art, design, technology and science, and a space to invite audiences to explore processes and participate. The Digital Design Weekend was going to include many more collaborative making sessions and hacks after that year.

The openFramerworks Lab had a great energy and the creative exchange continued long after the event ended. However, as all projects were initiated and led by either openFrameworks members or digital experts, despite the big number of audiences and interest, only few visitors had in-depth involvement or actually worked with the artists in any of the projects. If we wanted to engage visitors more in such events, we needed to find ways to break any barriers that stood in the way.

One element to review was our place and activity in the context of the Museum, but also the possibility of going beyond the four walls of the studio, the

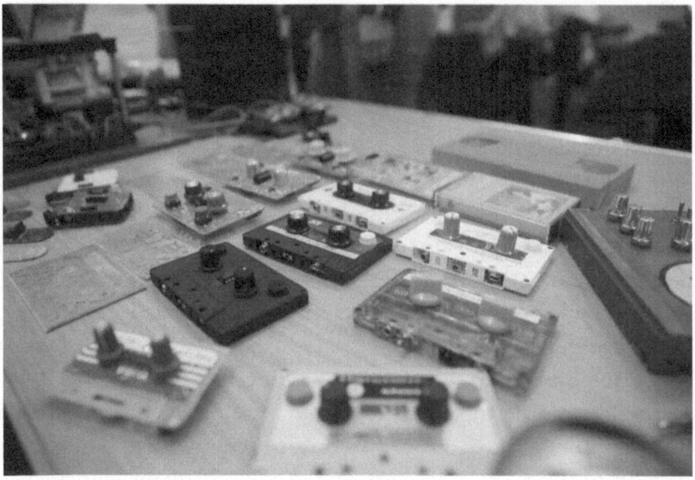

FIGURE 19.2 *8-bit Mixtape* by Andreas Siagian. Digital Design Weekend. © Victoria and Albert Museum.

main location of our events, and one of these barriers, since we were asking people to make a commitment by coming through the doors to engage in any of our programmes.

The opportunity for this first step came with the Power of Making exhibition and Tinkerspace,[7] where we temporarily moved from the studio a series of 'show & tell,' participatory activities with digital artists and designers. The exhibition presented making in our time, including beautifully crafted objects from both amateurs and leading makers; beading, blacksmithing, weaving, coding, rapid prototyping and 3D printing were a few of the techniques explored. Power of Making encouraged debate on the nature and importance of making, from problem solving and survival, but also as a means of learning and enjoyment. The Tinkerspace gave us a great chance to expose our work to curious visitors, who came by to join artists 'at work' while taking part in activities exploring processes and the use of technology in art and design.

Ele Carpenter's *Embroidered Digital Commons* artwork, as part of her Open Source Embroidery project,[8] was one of these first activities in the Tinkerspace; a distributed embroidery project that invited people to discuss the meaning of the term 'Vector' and the digital commons from Raks Media Collective's *The Concise Lexicon of/for the Digital Commons*,[9] while collectively hand stitching the definition of the term. The *Embroidered Digital Commons*, as well as other sessions that followed, experimented with an open and flexible format for engaging visitors with work that challenges our perception of digital and invited them to consider new technologies as craft and to draw parallels between digital work and process in objects from the collections. In a way we were exploring how through this open making or participatory work and discussion in the galleries we could 'hack' the traditional museum idea and explore new directions in public engagement.

The Tinkerspace left the Museum with the exhibition, but we decided to keep some programmes in the galleries, nomadic activities that could help create stronger links between contemporary digital art and design with current ideas. Such activities included projects that investigated big data and our perception of privacy, electronic waste and repair or repurposing to digital currencies, design and technology in health, surveillance and identity systems and more. These informal demonstrations and drop-in workshops were in a way a portal to start understanding digital society and start asking important questions. To quote Garnet Hertz, 'to be critically engaged with culture, history and society: after learning to use a 3D printer, making an LED blink or using an Arduino, then what?'[10]

When the time came to try out another hack event, this time we joined forces with an unlikely partner for the V&A; the Met Office. Being connected to scientists from the Met Office was one of these brilliant coincidences that turned into a long-term collaboration.

Back in 2012, I initiated an ongoing programme called Digital Futures.[11] The main reason behind this was the need for a regular and flexible platform not only to carry out research work but also to sustain collaborations developed through the annual Digital Design Weekend. An annual event was clearly not enough if we were to have a continuous and evolving relationship with partners, individuals and other organisations.

Digital Futures, which started as a series of informal meetups, mainly aiming to share work by academic researchers, PhD and postgraduate students with the public, soon developed into a networking event, a space for people to meet as well as a platform for bringing together not only different disciplines but also the industry. The format was open, experimental and responsive including showcases, discussions and workshops to long-term research projects.

It was through this programme that I was connected with Michael Saunby, a scientific software engineer from the Met Office, with whom we also shared a fascination for hackathons. Michael and his colleagues had been organising hackathons across the UK bringing together anyone from scientists and engineers to software developers and designers. I was fortunate to be involved as a judge in one of the most exciting of these, the NASA Space Apps Challenge,[12] from which we decided to share the outcomes in the Museum in order to allow a lot more people to experience how participatory design, art, technology and science can improve cities and life on earth, as well as life in space.

The showcase, which was part of a Digital Futures session in partnership with the Met Office, presented over 30 projects including accessible kits for growers, urban surveillance alert systems in the form of wearables and Lego rovers to engage kids with space exploration challenges.

We recognised that these hacks were exciting opportunities with incredibly creative outcomes, but we also felt there was a need to be more inclusive and diverse. Unfortunately, the hacks we were familiar with, although amazing events, were quite competitive and attracted mainly participants with computational and engineering skills, so it wasn't the format to engage wide audiences and sadly the majority of such events had very few women.

So, we decided to try a joint event with the Met Office and the Centre for Sustainable Fashion at the V&A; a Climate Change & Fashion Hackathon[13] using open collaboration to address environmental and social challenges arising from climate change. We invited climate scientists, artists, designers and Museum visitors to come together and prototype projects responding to a world adapting to changing climates. The hack involved mainly low tech and alongside 3D printers, climate data, Arduinos and Raspberry Pis, we included tools such as sewing kits and sewing machines, which created access points for people with no digital experience. The challenges also were accessible and relevant to general audiences since we looked at topics such as climate change and links with the fashion industry, pressures of an ageing population, urbanisation, bird migration patterns and more.

Participants explored solutions and ways to survive future extremes and variations, with a focus on adaptability and connectivity. At the end of the event, we had put together prototypes and new design methodologies to test ideas. The Climate Change & Fashion Hackathon was one of the most successful participatory events at the Digital Design Weekend, with a few hundred visitors taking part and many more coming through the doors to observe. But more importantly, these partnerships continued for a long time after the hack, with many of the people involved still part of the annual Digital Design Weekend or Digital Futures and similar activities.

With this event, we tried to re-imagine the existing structure of a hackathon and create a public open collaborative event to bring in citizens, artists, designers and technologists, but also the industry, and explore the museum as a site for collective innovation and production, for experimentation, dialogue and action.

As Jon Rogers says in his introduction to our publication for the Open Collaborative Making project that we organised a year later:

> Exploring values demands a highly participative approach. It needs people to be sharing knowledge, ideas and viewpoints together. To understand and harness the latent capacity within digital value, we need to do this in an open and collaborative way. In recent years hackathons have started to gain popularity as ways of bringing people together to solve problems in a shared space. I'm a little critical of what this means, because in the main this involves people being taken out of everyday life and everyday activities and placed in a closed space, often working through the night in secret teams. …I much prefer to think of this process as Collaborative Making, where it is more integrated into people's lives in a manageable way;[14]

The Open Collaborative Making project was an attempt to imagine what the digital future might be; not one that is being built without our vision or input, but a future that we want, a future that we can contribute to, shape and own. We invited visitors to explore opposite ideas such as an open, sharing Web against a closed internet where we become data mines serving the interests of few corporate machines; or to join workshops understanding and visualising data through physical objects and drawing, as well as building tools that help us perceive the amount of electronic waste generated in our homes. When it comes to imagining what the future might look like, we need to explore the importance of diversity, participation and focus on the human.

We have been regarding all this activity as an ongoing project, where we continue to learn and evolve ideas opening up new spaces and opportunities for exploration, exchange, collaboration, thinking and questioning. Keeping programmes such Digital Futures as a flexible, responsive and peripatetic platform has been important not only for the development of this work but also for reaching people beyond the Museum, such as collaborations with other cities and international organisations. One such example was an exchange and collaborative prototyping event between Mexico City, Dundee and London, on the themes of civic awareness and engagement and innovation, and inviting people to share ideas and work across the Atlantic responding to environmental and urban issues.[15]

Partnerships and collaborations have been central to all of these programmes, which have fostered national and international networks of participants from across a breadth of industries and sectors. We have been extremely fortunate to work with so many diverse groups and individuals, as each journey and event will take us in different directions through exploring ideas from completely different perspectives.

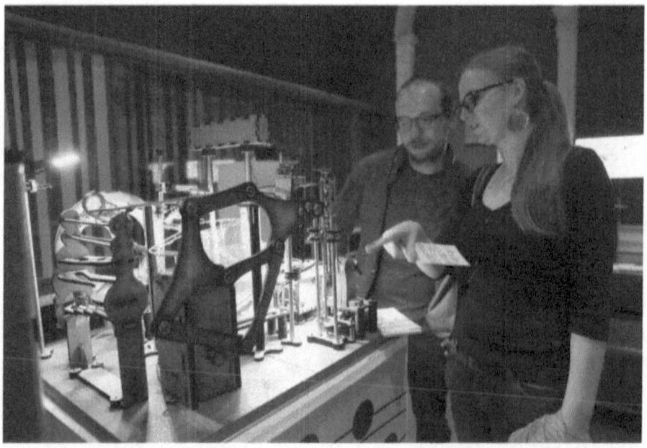

FIGURE 19.3 *Hola Goldberg!* by Centro de Cultura Digital, Mexico City. Digital Design Weekend. © Victoria and Albert Museum.

By this point, and when it comes to talking about our work, you will have noticed the repeated use of open, 'one of the keywords of the age.'[16] Open as in inclusive, malleable and resilient; open to receive different viewpoints and learn with others, as a forward-thinking organisation should be in relation to a field that is constantly changing. Thinking and acting along these terms, we set to open our doors to experimentation and collective work, opening up discussions about our future and setting the ground for activities that enable peer production and critical thinking.

> Openness isn't about untamed market economics. It isn't about unlimited consumer choice. It isn't even necessarily about freedom of movement. It is about access to knowledge. The concomitant of this is that an open society is a questioning society. It is a society that doesn't take anything on trust but wants to investigate and interrogate. It doesn't blindly follow precedent but wants to work things out for itself. It questions and interrogates what words like 'open', 'disruptive' and 'transformative' mean and how they are used.[17]

Notes

1 Ratto, Matt, and Stephen Hockema. 2009. "FLWR PWR – Tending the Walled Garden." In *Walled Garden*, Annet Dekker and Annette Wolfsberger (eds), Virtueel Platform, Amsterdam, 51–60.

2 In November 2006, Nina Simon started the Museum 2.0 blog, http://museumtwo.blogspot.co.uk, exploring how Web 2.0 can be applied in museums.

3 Sleigh, Andrew, Stewart, Hanna and Stokes, Kathleen, Open dataset of UK makerspaces, NESTA, 2015, https://www.nesta.org.uk/uk-makerspaces-data.

4 Hellicar & Lewis, formed by Pete Hellicar and Joel Gethin Lewis, was a craft, design and technology studio based in London between 2008 and 2016.

5 Decode: Digital Design Sensations was exhibited at the V&A from 8 December 2009 to 11 April 2010. The exhibition was divided in three areas including computational code as a design tool; interactivity and how it allows a range of relationships between viewer and artwork using sensors, tracking, cameras, feedback and coding; and finally networks, exploring how our networked world has provided artists with a range of tools.

6 The Digital Design Weekend started in September 2010 as a contemporary digital art and design annual festival at the V&A that coincided with London Design Festival. It still runs today.

7 Power of Making was exhibited at the V&A from 6 September 2011 to 2 January 2012, curated by Daniel Charny. The Tinkerspace was a participatory demonstration area within the exhibition space.

8 Open Source Embroidery is an ongoing socially engaged art project initiated in 2005, investigating embroidery and code as tools in participatory, open source production and distribution models. https://research.gold.ac.uk/3111/1/osembroidery.htm.

9 *The Concise Lexicon* was written by the Raks Media Collective in 2003 and published in the Sarai Reader 03: Shaping Technologies, The Sarai Programme, CSDS & The Waag Society/for Old and New Media, 2003, pp. 357–365, http://sarai.net/sarai-reader-03-shaping-technologies/. The lexicon is a poetic and informative A-Z of the language of the commons.

10 Introduction to Critical Making, a DIY zine that discussed hacking, making and electronics in a less sanitised way. http://www.conceptlab.com/criticalmaking/.
11 Digital Futures was launched in May 2012 and has been running since then at the V&A or at host organisations in London, the UK and internationally. http://www.vam.ac.uk/blog/tag/digital-futures.
12 Space Apps is NASA's annual, global hackathon and incubator innovation programme. The hackathon takes place over 48 hours in cities around the world. https://2017.spaceappschallenge.org.
13 The Climate Change & Fashion Hackathon took place in September 2013 as part of the annual Digital Design Weekend, in partnership with and supported by the Met Office, the Centre for Sustainable Fashion and RS Components.
14 Rogers, Jon. 2014. "We Don't Know What The Digital Future Is." *Open Collaborative Making, Uniform Communications Ltd*, 3–6. http://uniform.net/media/59959/Open-Collaborative-Making-a-Digital-Perspective.pdf.
15 Digital Futures UKMX took place from June to September 2015. A partnership between the V&A, British Council, Centro de Cultura Digital, Laboratorio para La Ciudad, Laboratorio Arte Alameda, Creative Dundee, Small Society Lab, Goldsmiths University and Dundee University. https://creativeconomy.britishcouncil.org/projects/digital-futures-ukmx/.
16 Prescott, Andrew. 2017. "What Do We Mean by 'Open.'" *Bridging Open Borders, Uniform Communications Ltd*, 49–51. https://digitransglasgow.github.io/bridgingopenborders/contributions/17_WhatIsOpen.html.
17 Ibid.

Bibliography

Anderson, Chris, 2013. *Makers: The New Industrial Revolution*, Random House Business Books.

Bridging Open Borders. 2017. Part of V&A Digital Design Weekend 2017, Uniform Communications Ltd, https://digitransglasgow.github.io/bridgingopenborders/.

Bridging Open Borders.

Charny, Daniel (ed). 2011. *Power of Making, The Importance of being Skilled*, V&A Publishing and the Crafts Council.

Crafting Our Digital Futures. 2015 Part of V&A Digital Design Weekend 2015, Uniform Communications Ltd, https://issuu.com/creativeeconomybritishcouncil/docs/crafting_our_digital_futures.

Damiani, Luca M. and Irini Papadimitriou. 2017. "MozEx: Web? Arts? Tech? Experience?", Electronic Visualisation and the Arts (EVA 2017) Conference Proceedings, July 2017, London, 11–13.

Decode: Digital Design Sensations exhibition guide. 2009. Victoria and Albert Museum.

Engineering the Future. 2016. Part of V&A Digital Design Weekend 2016, Uniform Communications Ltd, https://lguariento.github.io/Engineering-the-Future/.

Greenfield, Adam, 2017. *Radical Technologies*, Verso.

Hertz, Garnet. 2012. *Critical Making*, Telharmonium Press, http://www.conceptlab.com/criticalmaking/.

Hertz, Garnet. 2018. *Disobedient Electronics*, The Studio for Critical Making, http://www.disobedientelectronics.com.

Levy, Steven, 2010. *Hackers: Heroes of the Computer Revolution*, O'Reilly Media.

Open Collaborative Making. 2014. Part of V&A Digital Design Weekend 2014, Uniform Communications Ltd, http://uniform.net/media/59959/Open-Collaborative-Making-a-Digital-Perspective.pdf.

Ratto, Matt, and Stephen Hockema. 2009. "FLWR PWR – Tending the Walled Garden." In *Walled Garden,* Annet Dekker and Annette Wolfsberger (eds), Virtueel Platform, 51–60.

Sarai Reader 03: Shaping Technologies. 2003. The Sarai Programme, CSDS & The Waag Society/for Old and New Media.

Simon, Nina. 2010. *The Participatory Museum*. First Edition. Museum 2.0.

INDEX

Note: **Bold** page numbers refer to tables; *italic* page numbers refer to figures and page numbers followed by "n" denote endnotes.